inci
be
L

0

3(

POLLINATION
POWER

A purple-throated mountain-gem (*Lampornis calolaemus*) hovers to feed on purple porterweed (*Stachytarpheta frantzii*), Costa Rica.

POLLINATION
POWER

Kew Publishing
Royal Botanic Gardens, Kew

The University of Chicago Press
Chicago and London

Published in 2015 by the Royal Botanic Gardens, Kew,
Royal Botanic Gardens, Kew, Richmond, Surrey, TW9 3AB, UK
www.kew.org
and
Published in 2016 by The University of Chicago Press, Chicago 60637, USA

25 24 23 22 21 20 19 18 17 16   1 2 3 4 5

Kew Publishing ISBN 978 1 84246 606 3
The University of Chicago Press ISBN–13: 978-0-226-36691-3 (cloth)

British Library Cataloguing in Publication Data
A catalogue record for this book is available from the British Library.

Library of Congress Control Number: 2015948593

∞ This paper meets the requirements of ANSI/NISO Z39.48-1992 (Permanence of Paper).

Copy editing and proof reading: Ruth Linklater
Design, typesetting and page layout: Nicola Thompson, Culver Design
Production: Georgina Smith

For information or to purchase all Kew titles please visit shop.kew.org/kewbooksonline or email publishing@kew.org

Kew's mission is to inspire and deliver science-based plant conservation worldwide, enhancing the quality of life.
Kew receives half of its running costs from Government through the Department for Environment, Food and Rural Affairs
(Defra). All other funding needed to support Kew's vital work comes from members, foundations, donors and commercial
activities including book sales.

Printed and bound in Italy by Trento s.r.l.
Holder of the following quality, printing and environmental certifications:
ISO 12647-2:2004, Fogra® PSO (Process Standard Offset),
CERTIprint®, ISO 14001:2004, Carbon Trust Standard.

CONTENTS

All the flowers of all the tomorrows are in the seeds of today. *Indian Proverb*

A male orange-breasted sunbird (*Anthobaphes violacea*) feeds on *Erica mammosa*, Kirstenbosch NBG, Cape Town, South Africa.

INTRODUCTION

Rounding a corner of a byway in central Texas, a field — as blue as the ocean — came into view. On closer inspection, I saw the colour originated from wall-to-wall small blue lupines or bluebonnets, the Texas state flower. Unlike the azure carpets of bulbous bluebells that transform many British woodlands every spring, the bluebonnets are annuals. Such a landmark year as I encountered in Texas, depends on two essential factors: plenty of seed produced in the previous years and ample rain in the winter months to trigger successful seed germination. The bluebonnets are but one of many annual flowers that paint kaleidoscopes of colour in prairies and along the byway borders during April as seen opposite. Likewise, in South Africa's Namaqualand — providing winter rain falls at the right time — open ground is carpeted with annual yellow daisies as far as the eye can see.

It is only possible to mention a few of the floral hotspots and stunning flowers that I have been lucky enough to see and photograph in over four decades. My first long haul trip that concentrated solely on plants, was in 1975 trekking on horseback in Kashmir; led by Oleg Polunin, who wrote many books, including co-authoring *Flowers of the Himalaya*. It was also my introduction to choice alpines such as *Paraquilegia anemonoides*, which I have since seen in China and Kazakhstan. The elevated viewpoint from horseback, was good for spotting larger flowers further afield, whereas small ones could be easily overlooked. It did not take us long to realise you had to pull up your pony before or after spotting a choice specimen, otherwise by the time you dismounted, it could be history!

With a few exceptions, this book is not about plants in their environment; rather the aim was to get in close to reveal the key floral parts, which aid reproduction and to show precisely how and where pollen is transferred to particular visitors which then carry it to another flower. In the case of lilies, it may be on the feet of hoverflies or the wings of butterflies and moths; whereas proteas have pollen presenters that transfer pollen to the head feathers of nectar-seeking birds. Here are flowers of annuals and perennials as well as shrubs and trees that originate from temperate, subtropical and tropical regions. Many were photographed at Kew Gardens and in my own Surrey garden. However, wherever possible, floral visitors and pollinators were taken in their natural habitat within twenty countries. The rich floral kingdoms of the Cape in South Africa and Western Australia are every botanist's dream.

Left
Annual prairie wildflowers growing in a ditch include bluebonnets (*Lupinus texensis*) and Indian paintbrush (*Castilleja indivisa*) in a landmark year. Texas, USA.

It is fascinating to look backwards to pinpoint the factors that determined the route I chose to follow. My maternal grandmother triggered my enthusiasm for the natural world. During long summer holidays spent on my grandparents' Suffolk farm, she taught me the names of wildflowers. Later on, I developed my own interest in insects — notably freshwater life. It was a natural progression to study zoology. Whilst a student, I took up scuba diving, and participated in a diving expedition to Norway; where I used my first camera to record marine life. Back in the UK, during a tedious day analysing statistics — never my forté — I wrote an illustrated article on sea anemones. Suddenly, I had discovered a way to boost my limited grant and an outlet for my photography. Not long after, I swapped careers from marine biology to wildlife photography. So began a peripatetic life travelling to several overseas locations a year to initially capture wildlife on film, along with a few striking plants. A macro lens transformed the way I looked at plants and whatever I am doing, I aim to shoot at least one new macro shot every day.

Photographs of plants and flowers are generally not such great sellers as images of cute animals or dynamic action, so it is rare for commissions to be offered to document plants far from home. I have been lucky enough to have had two projects.

The first originated when the editor of *Gardenia* magazine (the horticultural equivalent of the *National Geographic* with stunning photography) called from Milan in Italy to invite me to document the vegetation on the Rwenzori Mountains in Uganda. At this stage, I knew this area was famed for endemic giant lobelias and groundsels. Like Mount Elgon spanning the Uganda/Kenya border and Mount Kenya, these mountains resemble islands in the sky separated by elevation instead of surrounding sea. Here, over time, plants evolved as they adapted to altitude and aspect to become the endemic species we know today. Without hesitation, I accepted the brief — although the trip turned out to be the toughest I have ever made. Icy rivers had to be crossed by wading up to the waist; while a bog at over 3,000 metres elevation was essentially a patchwork of high hard tussocks emerging from boggy ground. So whether you squelched through the bog or leapt in a zigzag fashion from one tussock to another, ankles were constantly bashed. Also, the porter delegated to carry my gear was never in sight when I needed to work, so I ended up carrying it all myself. It was hardly surprising I lost half a stone in ten days.

Left
A flowering giant lobelia (*Lobelia deckenii*) with tree heath and *Helichrysum* sp. near Kabamba, Rwenzori Mountains National Park, Uganda.

Nonetheless, flowering glory lilies (*Gloriosa superba*) scrambling over shrubs beside the tracks, *Disa* orchids and outsized tree heaths, festooned with extensive epiphytic mosses, spurred me on to the first giant lobelias, where I had a fleeting glimpse of a sunbird sipping the nectar. At this stage, my passion for observing and photographing known pollinators in action had not been aroused and with a brief to complete, I could not linger.

A decade later, the British Council invited me to document the biodiversity of the Himalaya — in one month! The timing was perfect for the flowering of the tree-like rhododendron known as burans (*Rhododendron arboreum*), with an assortment of birds visiting the red flowers. It was during this trip the germ of an idea for recording examples showing the co-dependence of animals and plants originated.

A special interest in pollinators developed from watching them at work in my own garden and overseas. The diverse Living Collection at Kew Gardens inspired an ambitious five-year project looking at floral structure and pollinators, both in Britain and all over the world. This book is a taster of what will be a much meatier book with many more illustrations depicting my own discoveries as well as piecing together pollination research from across the world on specific species.

My favourite haunts in China are Sichuan and Yunnan — much travelled and explored by the plant hunters, who brought to the West many of the plants we now treasure in our gardens. E. H. Wilson introduced many plants from China and, based on information on a specimen label in Kew Herbarium, he travelled 1,000 km in Sichuan to collect seed of the striking red poppywort (*Meconopsis punicea*). This plant was a target species on one of my Sichuan trips, and the first specimen found near Balang Shan pass, was bejewelled with raindrops. Whenever I gently uplifted the petals, I found either flies or hoverflies inside, which tend to be more active at cooler and higher altitudes than bees. Wilson also introduced the regal lily (*Lilium regale*), that now graces many a herbaceous border in Western gardens. After Wilson found them growing on slopes overlooking the Min River in Sichuan, he wrote 'In the Min Valley the charming *L. regale* luxuriates in rocky crevices, sun-baked throughout the greater part of the year'. Today, looking up from the road in this location, the lilies are still clearly visible clustered in the crevices, although reaching them with a camera and tripod is not for the faint-hearted where 45° slopes are covered with loose shale.

Right
Raindrops on red poppywort (*Meconopsis punicea*) near
Balang Shan pass, is one of several different coloured
Chinese *Meconopsis* species. Sichuan, China.

The easiest flowers to photograph are those grown in my own garden, allowing daily checks to capture them in their prime condition. The hardest took me four years to get any photos at all. In 2010, during a tour of the Tropical Nursery at Kew — where some 34,000 tropical and temperate plants are kept and propagated behind the scenes, in 21 different climatic zones — I was shown plants of the orchid *Bulbophyllum echinolabium*, which was not flowering at the time. For the three successive years, I failed to see the flower: one year the bud dropped after it was watered, another I was abroad when it flowered and it failed to flower one year. So, by 2014, I had given up all hope, when I was looking at other orchids and overheard it was now flowering. It appears on page 118.

Time spent watching flowers will be rewarded by visitors coming and going, some feeding on nectar, some on pollen — the two most common rewards flowers offer their pollinators . However, by no means all visitors are pollinators. Painstaking fieldwork is necessary before a visitor can be confirmed as a pollinator, therefore, throughout this book, floral visitors are referred to as pollinators only when this has been proven by published work. The images of visitors include more than the familiar bees. They range from beetles, flies, hoverflies and bee flies, wasps and hornets, to moths and butterflies as well as a gecko, birds and mammals.

Flowers on trees are rarely as accessible to view and photograph as those of annuals and biennials, but binoculars and a telephoto lens help to bring them closer. Open grown trees, with plenty of light, in spacious gardens or parks, are more likely to produce flowers on lower branches. Tropical vines with flowers on pendulous stems that reach at least to eye level and below, are very accessible. Originating from the Philippines, the jade vine (*Strongylodon macrobotrys*) is always a star attraction when the large trusses of turquoise flowers appear in a tropical glasshouse. They are reputedly pollinated by a fruit bat. Whilst combing the internet for images of visitors to other flowers and emailing several people, I discovered that sunbirds also visit for the nectar; both in the Philippines and in Queensland, Australia where the vine grows outside and the olive-backed sunbird (*Nectarinia jugularis*) is native. Research is needed to see whether both bats and the sunbirds pollinate the flowers as they do with other leguminous tropical climbers.

Right
The luminous turquoise-coloured flowers of the jade vine (*Strongylodon macrobotrys*) arise from two pigments in a slightly alkaline cell sap. Kew Gardens, Surrey, UK.

Thanks to digital photography, it was possible to use special techniques for some images. Most notably focus stacking (by taking a series of focus slices and combining them using a software program to gain maximum depth of field for three-dimensional flowers), high speed flash (to freeze simulated buzz pollination) and a special ultraviolet flash to reveal UV guides hidden to our eyes as well as pinpointing exposed nectar resources. Additional information about the photography appears in the Photographic Notes.

Anyone who lives in a temperate zone and keeps a phenological calendar of when certain plants start to flower each year, be it in your own garden or a local nature reserve, will be well aware how extreme weather affects the time when flowers open. If there is a prolonged winter followed by a late spring, or a mild winter and an early spring, the time when plants flower may no longer synchronise with the emergence time of their insect pollinators.

The foraging activities of insect pollinators, in particular, are restricted by both wind and rain since they cannot fly in these conditions. Birds on the other hand, are able to fly and visit flowers in light wind and rain that is not too heavy, so their pollinating efforts are curtailed less by weather conditions than those of insects. Birds, as well as bats, fly faster than insects, carrying pollen over much greater distances, thereby enhancing the gene pool via cross-pollination.

Town and city gardens, regardless of their size, can help to increase resources for urban pollinators. In addition to planting pollinator friendly plants that provide readily accessible pollen and nectar, bee hotels increase solitary bee populations and food plants of butterflies and moths will encourage these adult insects to visit the garden. This book aims to bring pollination mechanisms to life so that readers will be encouraged to take a closer look at flowers, and their visitors, at a time when pollinators are declining globally.

Heather Angel

Right
Once the sun reaches carpets of *Crocus korolkowii* in snow melt areas, solitary bees and bumblebees arrive to forage, near Ansob Pass, Tajikistan.

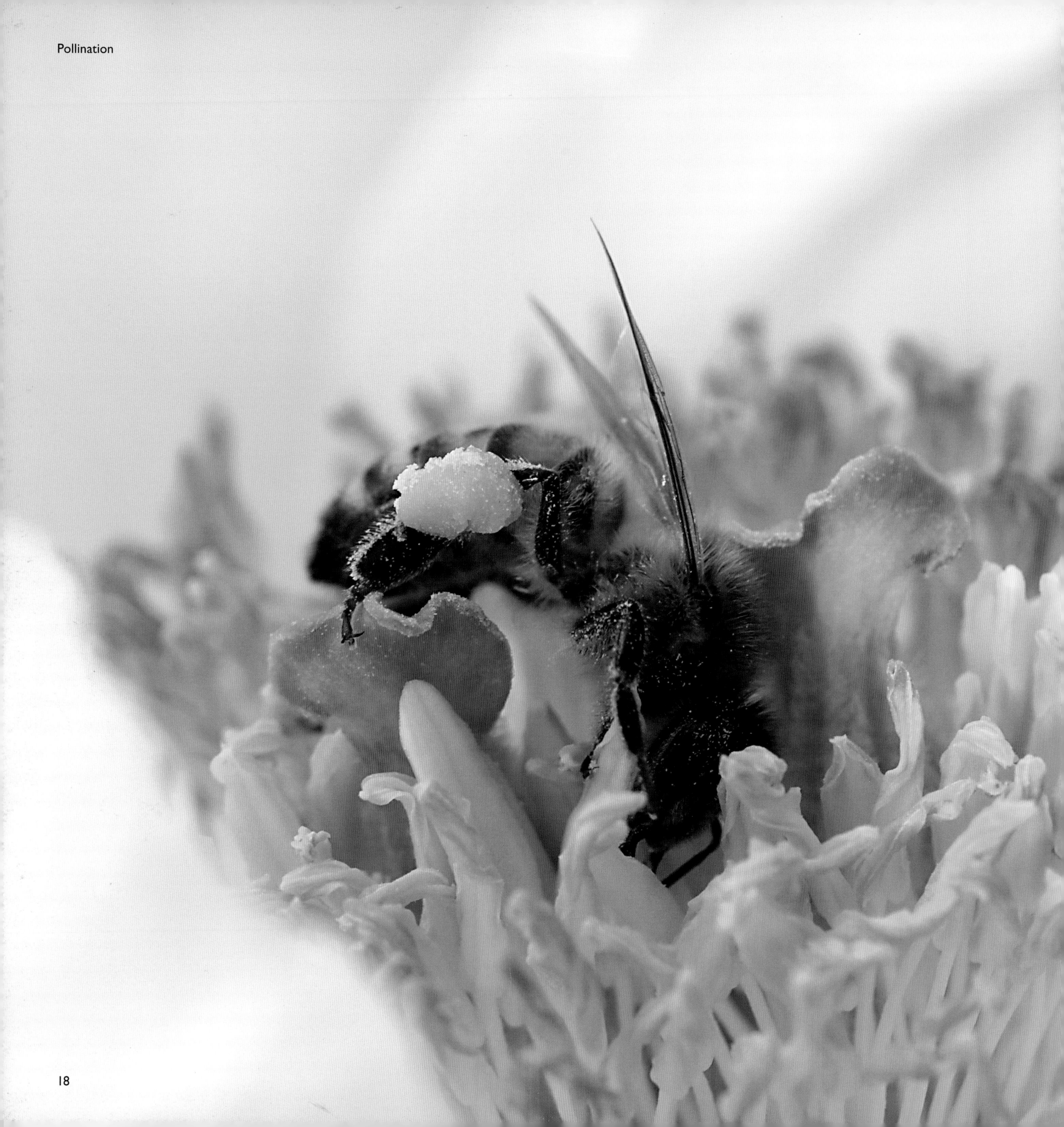

CHAPTER 1

POLLINATION

Flowers produce seeds after pollination takes place, when pollen is transferred to the stigma. Wind and water both play a part, but most plants rely on animals to deliver their precious pollen. New types of pollinators are being discovered — both by day and at night. As well as bats, the list of non-flying mammals continues to grow from small rodents in South Africa to marsupials in Australia.

A honeybee forages on pollen in a Caucasian peony (*Paeonia daurica* subsp. *mlokosewitschii*). Inside the wide ring of stamens, pollen has been deposited onto the erect pink stigmas. Kew Gardens, Surrey, UK.

Compared to animals, some plants have an extraordinary lifespan — more than 5,000 years for Great Basin bristlecone pines (*Pinus longaeva*). By then, such trees are misshapen from buffeting by strong winds and fungal rot, resulting in broken or lost branches to create unique and often highly characterful trees. Regardless of their age, plants like animals, do need to reproduce to guarantee another generation succeeds them. Over three quarters of all plants rely on animals to transport their pollen.

Just because an animal is seen visiting a flower however, it may not be the pollinator. This is how pollination myths arise. One example is that sunbirds pollinate the extraordinary sculptural bird of paradise (*Strelitzia reginae*) flowers that emerge one at a time from a stout boat-shaped spathe. Sunbirds do alight on the spathe rim to reach the nectar, but in reality these birds are nectar thieves since they fail to make contact with either the anthers or the stigma.

White sticky pollen, hidden inside the blue anther sheath, is revealed when a heavier weaver bird alights on the sheath, exerting pressure on the blue anther flaps that open along a central longitudinal slit. The image opposite shows how a weaver bird picks up the sticky pollen on its feet as it alights and also on its bill as it bends down to feed on the pollen. For pollination to occur, the pollen has to be transferred to the elongated white stigma that emerges from the end of the anther sheath. A description of spotted-backed weavers (*Ploceus cucullatus*) feeding in this way in Kirstenbosch National Botanic Garden in Cape Town, was published four decades ago but somehow has been overlooked — possibly by birders recording sunbirds feeding on nectar and assuming they were pollinating in the process.

Asexual reproduction

Some plants are able to reproduce asexually without producing any flowers or pollen being exchanged. Quite simply, new daughter colonies are budded off inside the spherical freshwater green alga *Volvox*. A single mother of thousands *(Kalanchoe daigremontiana)* plant can produce hundreds of plantlets along the edge of the succulent leaves, that drop to the ground, take root and grow into new plants. Strawberries reproduce asexually by sending out runners on the ends of which a little plant develops.

Sexual reproduction

The majority of flowering plants reproduce sexually and this is done by pollen grains being transferred to the stigma of the same plant (self-pollination) or another plant of the same species (cross-pollination). The gene pool is enhanced by cross-pollination combining a new assortment of genes from both male and female parents, thereby increasing the odds of fitter offspring being produced that are better adapted to changing conditions.

Left
A female southern masked weaver (*Ploceus velatus*) perches on the blue petalline sheath
of bird of paradise (*Strelitzia reginae*). The bird's feet have picked up white sticky pollen
and some remains on the bill after feeding. Pretoria, South Africa. © Ruslou Koorts.

Self-pollination is comparatively rare, but does occur in some orchids with friable pollinia that fragment, shedding pollen onto the stigma. In addition, plants that grow in extreme climatic conditions where pollinators are rare, can radiate out with seeds produced via self-pollination. Should cross-pollination fail, some flowers self-pollinate as a back-up. The marvel of Peru (*Mirabilis jalapa*) is pollinated by hawk-moths, but the moths fail to fly on cold nights. Then the next morning, the flowers self-pollinate as the floral lobes and floral parts inroll transferring pollen to the stigma.

Plants that are typically cross-pollinated, have devices that inhibit self-pollination such as the male and female flowers developing on separate plants. These dioecious (two homes) plants include holly (*Ilex aquifolium*) which needs to have a male tree nearby a female one for the latter to produce fruits. Monoecious (one home) plants have either bisexual flowers or both male and female flowers on the same plant. Many bisexual flowers have a mechanism for preventing self-pollination; namely, the male anthers mature and shed their pollen either before the stigma is receptive (protandrous flowers) or after the stigma is receptive (protogynous flowers). Roses and lilies are examples of bisexual flowers, while pumpkins and marrows have separate male and female flowers on the same plant.

A pin-eyed bogbean flower shows the stigma at a higher level than the stamens. The fringed hairs on top of the petals protect the nectar from rain. Farnham, Surrey, UK.

Bogbean with thrum-eyed flowers in which the stamens appear at a higher level than the stigma inside the top of the corolla tube.

Self-pollination is also prevented in flowers where the anthers and stigma lie at different levels. Bogbean (*Menyanthes trifoliata*) has two flower morphs: pin-eyed ones have a long style standing proud of short stamens, while thrum-eyed ones have long stamens above a short stigma in the corolla throat. Flowers with short stamens have their pollen transferred via visiting bumblebees so they pollinate flowers with a short stigma. Hence, pollen from long stamens is precisely picked up so that it only pollinates flowers with long stigmas. Bogbean flowers are more open than primroses (*Primula vulgaris*) — which also have pin and thrum-eyed flowers — so the distance separating the stigma from the stamens is clearly visible without having to the dissect flower.

The male cells are contained in pollen grains, which need to be transferred to another plant of the same species for cross-pollination to take place. However, since plants are fixed, they are dependent on agents to act as their pollen vectors. They have to rely on water, wind or animals to carry their precious pollen produced by the male sexual parts of the flower — the anthers on the stamens — to the female part (the stigma) of another flower. Pollination by water is restricted to a few aquatic plants and is comparatively rare. Known as hydrophily, the pollen of some aquatic plants floats on the water surface where it is carried to the female parts of other flowers by water currents or wind rippling the surface of still water. Many more plants rely on water — both freshwater and the sea — to disperse their seeds and fruits.

Wind pollination, known as an anemophily, is much more widespread and includes most coniferous trees — such as pines, yews, cedars and spruces — as well as some 12% of all flowering plants, including birch and walnut trees, which both produce catkins; grasses as well as cereal crops. Since this type of pollination is such a chancy affair, copious amounts of small and light pollen grains are liberated from small inconspicuous flowers that may lack petals and have no need to produce nectar. The stamens are exposed rather than enclosed within petals so that the pollen is freely shed and the feathery stigmas are quite conspicuous for easy pollen pick-up.

 If a branch of a coniferous tree with open male flowers is shaken on a warm, dry day, pollen appears as a yellow cloud. Anyone who suffers from hay fever, should not try this without wearing a mask. Much of the pollen released from conifers is wasted and even a single tree creates an ephemeral yellow scum on adjacent puddles or ponds.

On the other hand, since animal pollinators carry pollen directly to a flower, less pollen is wasted than when wafted around by unpredictable air currents. Therefore, smaller quantities of pollen can be produced and many flowers direct their energy into producing nectar as another reward and often produce showy flowers to attract their pollinators. Pollen grains dispersed by animals are larger than those of wind pollinated plants, rough and sticky so they adhere to the animal vector.

Insect pollinators

Since insects are the most abundant animal group, it is hardly surprising that insect pollinators are more numerous than any other animals. Honeybees and bumblebees maybe the best known and easily recognised, but the importance of solitary bees as pollinators has been underestimated. Unlike honeybees, bumblebees and wasps, which are social insects, adult solitary bees meet and mate on their favourite flowers. The female then constructs a nest with brood chambers provisioned with pollen fodder for the larvae when they emerge.

Honeybees, bumblebees and orchid bees collect pollen and transfer it to the pollen baskets on the hind legs. Other bees, as well as flies, hoverflies and wasps, carry pollen on their hairy bodies. Birds carry pollen on their feathers, bills and even on their feet, while mammals carry it usually on their muzzle fur as they reach inside a flower for the nectar. Perhaps the most bizarre place for pollen pick-up is on the tip of the tongue of lesser double-collared sunbirds (*Cinnyris chalybeus*) when they feed on the red, scentless flowers of the climber *Microloma sagittatum* in South Africa. No insects visit the flowers that fail to fully open, but pollen is transferred in pollen packed sacs known as pollinaria.

A bumblebee emerges from a squash flower covered
with large pollen grains. Kew Gardens, Surrey, UK.

Beetles are the most prolific insect group and they pollinate many more flowering plants than any other invertebrate pollinator. These insects are important pollinators in tropical and arid areas and in South American cloud forests they make up 45% of all pollinators. Many use flowers as places to meet and mate. Detailed studies on some beetle-pollinated flowers, reveal beetles are not the primitive pollinators once thought and their visitations to some flowers are controlled precisely by the flower itself. For example, *Magnolia grandiflora* attracts beetles by emitting a lemony scent specifically when the female phase is receptive. The flowers then close, trapping the insects overnight and reopen the following morning after the anthers have dehisced, so the beetles depart carrying a pollen load.

As a hoverfly feeds on a grape hyacinth (*Muscari neglectum*) out of the wind in the Davies Alpine House at Kew, white pollen falls down onto the insect. Kew Gardens, Surrey, UK.

The activity of some pollinators is highly weather dependent. The mobility of insects in particular is governed by weather, especially the air temperature. Flies tend to become active when the temperature is several degrees below 10°C; whereas honeybees need at least 10°C — sometimes more — before they start to forage. Rain and wind also tend to restrict flying insects whereas birds can fly in light rain, providing the wind is not too strong. Early in spring, hoverflies, honeybees and bumblebees have discovered there is a wealth of rewards to be gained by foraging inside the sheltered Davies Alpine House at Kew.

Bee-flies have a special association with primroses and are their main pollinators. They superficially resemble bees, but a closer look reveals a long proboscis, which remains extended as they fly in a strange staccato zigzag fashion fairly close to the ground. These insects require a minimum temperature of 17°C before they are active. Therefore, early on in the day they spend more time sunbathing than foraging. However, once they warm up, they either rest on the flower to feed (p42) or hover with some legs briefly touching the flower (p43).

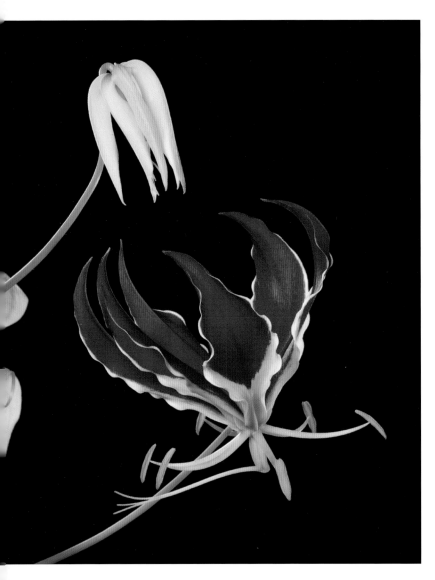

Within an open bowl flower such as a tulip, an anemone or a wild rock rose, all of the reproductive parts are clearly visible and readily accessible to insects. Yet, the red-bowled 'beetle' flowers from the Mediterranean region with black marks on the petal bases of the tulips or the black stamens of anemones, selectively attract scarab beetles which use the flowers as places to meet and mate. Insect visitors to any open bowl flower deposit pollen, carried from another flower as they crawl around the bowl and also pick up a new pollen load.

As well as open bowls, bees also visit flowers with an entrance via a slit between an upper and lower lip, which they prise open. The proboscis length of each species determines which flowers they can reach into to extract nectar. Butterflies and moths, each with their long narrow proboscis are able to extract nectar from narrow tubular flowers.

The glory lily (*Gloriosa superba*) flower is designed for butterfly pollination, with six reproduction units, each with one stamen, but only one has a style. This

Individual stalked glory lily flowers open in succession from a stem with the green buds gradually turning to red and yellow to attract pollinating butterflies. Farnham, Surrey, UK.

makes a right-angled turn so the stigma projects towards the brightest light from where butterflies typically approach the flower. As they feed on nectar, the wings flutter against the stamens, picking up pollen on the underwings as butterflies do when feeding on hibiscus flowers shown on page 106.

Fewer moths are active by day than by night, but some, such as burnet moths pick up pollen (collected together in two small packets known as pollinia) on the proboscis when feeding on wild orchid flowers. Nocturnal moths — notably hawk-moths — are important nocturnal pollinators, but since our eyes are not well adapted for night vision, it is much more difficult to study nocturnal animals.

When Charles Darwin saw the lengthy spur on the comet orchid *Angraecum sesquipedale* from Madagascar amongst a collection of orchids he received, he predicted the pollinator would have to have a proboscis almost as long as the 29cm spur to reach the nectar. After Darwin died, the African hawk-moth (*Xanthopan morganii praedicta*) was discovered in Madagascar and the orchid is now referred to as Darwin's orchid.

The waxy bloom of Darwin's orchid (*Angraecum sesquipedale*) with the lengthy spur, which is visited by the hawk-moth, *Xanthopan morganii praedicta* and is assumed to pollinate the orchid. Kew Gardens, Surrey, UK.

Vertebrate pollinators

Pollination ecologists are discovering new types of pollinators — by day and night — including some now extinct, using unconventional techniques to unravel past plant-pollinator associations. Amongst vertebrates, lizards and geckos are known to pollinate some native flowers on Atlantic and Indian Ocean Islands. Typical of island flora and fauna, endemic species have evolved to become better adapted to local conditions. On Madeira, the endemic golden bellflower (*Musschia aurea*), shows secondary pollen presentation on the robust emergent stigma before it opens out ready for pollination by lizards and bumblebees.

Birds

Nectar seeking neotropical hummingbirds, old world sunbirds and Australasian honeyeaters are well known as floral visitors which pollinate flowers in the process. Today, the Hawaiian honeycreepers are either extinct or endangered. When plants are transported from one garden to another across continents, birds adapted for extracting nectar soon discover a new nectar source to exploit and gardeners living in both North and South

The golden bellflower (*Musschia aurea*) has a stigma and style robust enough to resist the weight of lizards as they feed on nectar and pollinate the flower in Madeira. Kew Gardens, Surrey, UK.

America enjoy hummingbirds visiting nectar-rich exotic flowers grown in their gardens. Nectar robbers also visit to extract nectar but do not enter via the legitimate route and so they fail to pick up or deposit pollen.

Hummingbirds with long bills, such as the Andean sword-billed hummingbird (*Ensifera ensifera*), can reach the nectar in flowers with long narrow corolla tubes as in *Passiflora mixta*. Most often, they hover to feed, but I have seen hummingbirds repeatedly alight on the long spathe-like primary bract of the expanded lobsterclaw (*Heliconia latispatha*).

Some trees attract an eclectic range of visitors when they flower. The drunken parrot tree (*Schotia brachypetala*) from South Africa is one of these that has to date a total of 54 different kinds of birds recorded as feasting on the copious nectar. Any nectar which remains uneaten, ferments after a few days causing a mild narcotic effect. This illustrates the difficulty of identifying the true pollinators amongst so many visitors. Eleven of these are sunbirds, which are the most likely candidates.

A rainbow lorikeet (*Trichoglossus moluccanus*) feeds on nectar from a drunken parrot tree (*Schotia brachypetala*). Royal Botanic Gardens, Sydney, Australia.

Thanks to the pollen of animal pollinated flowers being virtually distinct for each species, pollen found on clothing has been used to place a person at a crime scene. Pollen has also solved the mystery of historic pollinators of the Hawaiian liana *Freycinetia arborea*, known locally as the *ieie*. Today, the rewarding red sugary bracts are eaten by the Japanese white-eye (*Zosterops japonicus*), which was introduced from Japan in 1929. To discover the native pollinators, searches in the dairies and notebooks of early Hawaiian naturalists unearthed the names of birds seen feeding on the red bracts. Using museum specimens, feathers examined under an electron scanning microscope revealed *ieie* pollen loads on two Hawaiian honeycreepers: the o'u (*Psittirostra psittacea*) now critically endangered and the extinct Kona grosbeak (*Chloridops kona*), with no trace of pollen from any other flowers.

Mammals

Bats are important pollinators of flowers at night, both in the Neotropics as well as the Old World and in Australasia. Bat-pollinated flowers tend to open at dusk or during the night, have dull colours and emit a musty odour. The entrance to the flower is wide and large amounts of both nectar and pollen are produced.

A honey possum feeds on nectar and pollen from a Stirling Range bottlebrush (*Beaufortia cyrtodonta*) after retrieval from a pit trap early in the morning. Cheyne Beach, Western Australia. © Stephen D. Hopper.

In the 1970's a new type of mammal — the Namaqua rock mouse (*Aethomys namaquensis*) — was discovered to pollinate a protea (*Protea humiflora*) with musty smelling flowers close to the ground in South Africa. Since then, several species of gerbils, mice, rats and shrews have been added to the list of non-flying mammalian pollinators (to distinguish them from bats). In Australia, small nocturnal or crepuscular marsupials, including the feathertail glider (*Acrobates pygmaeus*), the honey possum (*Tarsipes rostratus*) and the western pygmy possum (*Cercartetus concinnus*) feast on nectar of banksias, grevilleas and bottlebrushes (*Callistemon*) and pollinate the flowers.

The honey possum is one of a few mammals that feed entirely on nectar and pollen. It needs year-round nectar to survive and whilst feeding, it uses the front and hind feet as well as a long prehensile tail, to grasp shrubs. It is among several small marsupials in Australia that feed on flowers belonging to the protea and myrtle families. On average, a 9g animal drinks 7ml of nectar and eats 1g of pollen each day. For such a tiny animal it is remarkable the elaborate system involved in the collection and digestion of the pollen. Two front teeth in the lower jaw support the long tongue as it moves in and out of the mouth at rate of four times a second, lapping up nectar. Papillae on the surface aid pollen collection, which is scraped off the tongue by combs in the roof of the mouth. The contents of the pollen grains are digested, since only empty shells appear in the droppings.

Previously thought to be chiefly nocturnal, the honey possum is now known to be crepuscular, which reduces the risk of predation from both diurnal and nocturnal predators, since their eyes are not adapted for crepuscular vision.

To prove an animal visitor is a true pollinator, requires painstaking fieldwork observing pollen being transferred to a marked individual, marking pollen with a fluorescent dye so its passage can be tracked using an ultra violet light from the flower to the pollinator and then to another flower to confirm cross-pollination has taken place. Pollen picked up by nocturnal mammals is verified by the retrieval of animals from pit traps or live trapping and using sticky tape to collect pollen samples from the fur. Even if pollen is successfully transferred to the stigma, there is no guarantee that seeds will result. Sometimes pollen tubes fail to convey the male cells in the pollen grains to the ovary so they can combine with the female cells.

Next time you plant flower and vegetable seeds, pause for a moment to consider the work done by pollinators for flowers to produce their seeds. This book highlights pollination by animals — from minute insects to birds and mammals.

WIND POLLINATION
Female and male flowers of European larch
(*Larix decidua*) appear on the same branch just
as new needles begin to emerge. Male flowers
are clusters of golden stamens that release
copious pollen carried by wind to the erect
larch 'roses', Farnham, Surrey, UK.

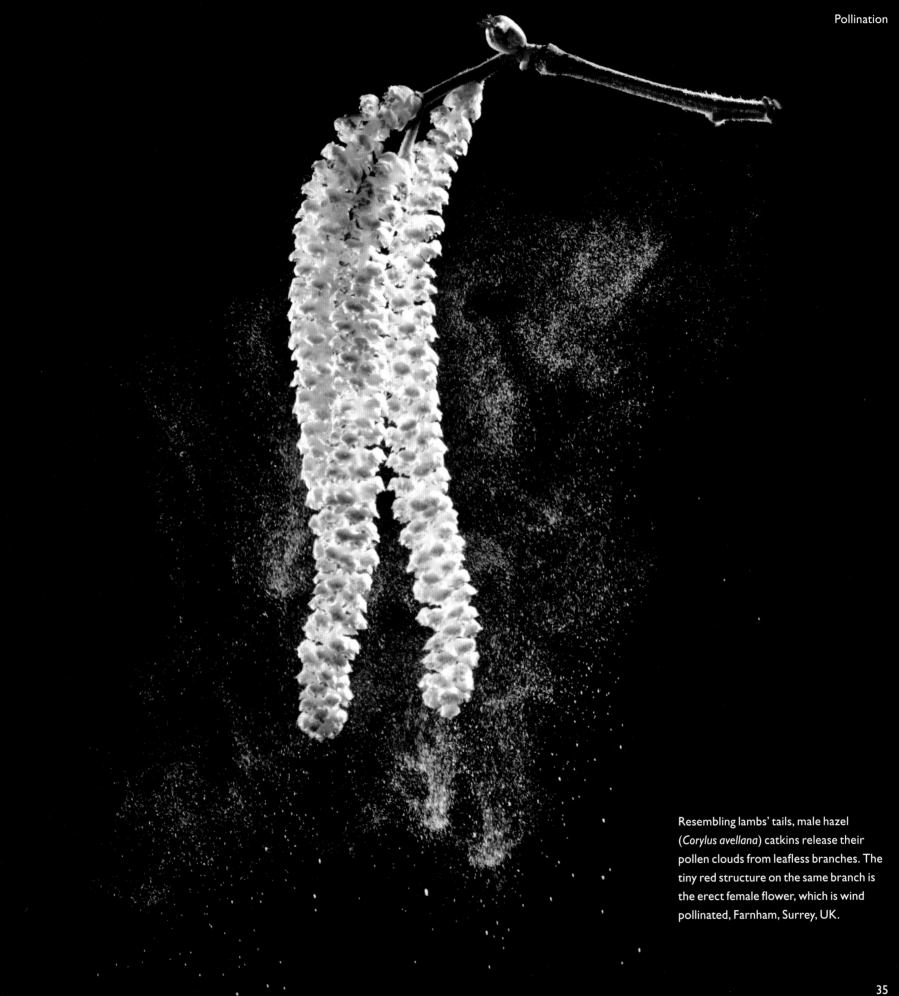

Resembling lambs' tails, male hazel
(*Corylus avellana*) catkins release their
pollen clouds from leafless branches. The
tiny red structure on the same branch is
the erect female flower, which is wind
pollinated, Farnham, Surrey, UK.

BEETLES

Above

In the eastern Mediterranean region, scarab beetles are attracted to red bowl flowers — including tulips where they meet and mate, picking up pollen in the process. These bumblebee scarabs (*Amphicoma vulpes*) are mating inside *Tulipa armena* near Karabet Pass, Turkey.

Right

Brightly coloured blister beetles (*Mylabris quadripunctata*) are conspicuous visitors to flowers, picking up pollen on their hairy bodies. However, they can cause damage to flowers by eating the petals as shown with this *Tragopogon* in Kazakhstan.

HOVERFLIES

A hoverfly wraps its tongue around a stamen to feed on viper's bugloss (*Echium vulgare*) pollen, Alpes Maritimes, France.

When a marmalade hoverfly (*Episyrphus balteatus*) rests on a 'Stargazer' lily anther to feed on pollen, some sticks to the legs. Then, if the insect lands on the stigma of another lily flower, pollen is transferred, Farnham, Surrey, UK.

BEE-FLIES

Left

A large bee-fly (*Bombylius major*) rests on a pin-eyed primrose (*Primula veris*) to sip nectar with its long proboscis, Farnham, Surrey, UK.

Below

Bee-flies also hover to feed on nectar. As it withdraws from an honesty (*Lunaria annua*) flower, the proboscis is coated in pollen. These insects fly with their proboscis outstretched, so the pollen can be transferred to another flower, Farnham, Surrey, UK.

SOLITARY BEES

Below
A female hairy legged mining bee (*Dasypoda hirtipes*) collects pollen from ragwort (*Jacobaea vulgaris*) where male bees cruise around looking for a mate, Farnham, Surrey, UK.

Right
As a female wool-carder bee (*Anthidium manicatum*) enters a rusty foxglove flower (*Digitalis ferruginea*) the anthers deposit pollen onto her head and thorax, Kew Gardens, Surrey, UK.

BUMBLEBEES

Above

A queen buff-tailed bumblebee (*Bombus terrestris*)

Right

Busily foraging on nectar from flowering rush

HONEYBEES

Left
Honeybees (*Apis mellifera*) are important
pollinators of many fruits we eat. This one is
collecting pollen from apple (*Malus domestica*)
blossom, Farnham, Surrey, UK.

Above
On a sunny early spring day, a
honeybee is smothered in crocus
pollen, which it collects to take
back to the hive, Surrey, UK.

WASPS AND HORNETS
Ivy flowers, (*Hedera* sp.) provide a valuable autumn nectar source for many insects including hornets (*Vespa crabro*), Kew Gardens, Surrey, UK.

Whilst feeding on nectar of a pineapple lily (*Eucomis comosa*), a social wasp (*Vespula germanica*) picks up pollen as it makes contact with the anthers and may deposit it on the stigma of another flower, Kew Gardens, Surrey, UK.

BUTTERFLIES
A Julia heliconian (*Dryas iulia*)
nectaring on a zinnia,
Chiapas, Mexico.

A cardinal fritillary (*Argynnis pandora*)
feeds from a thistle in Kazakhstan
alongside a blister beetle.

A yellow-legged tortoiseshell (*Nymphalis xanthomelas*) feeds on *Eremurus altaicus* in Kazakhstan.

REPTILES
A day gecko (*Phelsuma* sp.) laps up
Euphorbia nectar, Madagascar.

Male malachite sunbird (*Nectarinia famosa*) feeds on king protea (*Protea cynaroides*) — the national flower of South Africa.

When a male Cape sugarbird (*Promerops cafer*) bends down to feed on nectar from a pincushion protea (*Leucospermum cordifolium*), the head feathers pick up pollen from the pollen presenters, the Cape, South Africa.

A Cape weaver (*Ploceus capensis*)
blends in with a flowering *Aloe*, as it
feeds on nectar, the Cape, South Africa.

MAMMALS

As a Geoffroy's tail-less bat (*Anoura geoffroyi*) inserts its head into a *Cobaea trianae* bell to feed on nectar, pollen is transferred to the throat and chest, Pichincha Province, Ecuador.
© Nathan Muchhala / www.umsl.edu/~muchhalan

A honey possum (*Tarsipes rostratus*) feeds on scarlet
banksia (*Banksia coccinea*), with the pollen presenters
providing the flower colour. Western Australia.
© Gerry Ellis / FLPA

Right
At night, a mother of pearl moth
(*Pleuroptya ruralis*) visits ragwort to
feed on nectar, Farnham, Surrey, UK.

DAY AND NIGHT FLOWERS
Above
During the day, red soldier beetles (*Rhagonycha fulva*)
meet and mate on ragwort (*Jacobaea vulgaris*) and pick
up pollen on their bodies. Farnham, Surrey, UK.

NOCTURNAL FLOWERS

Evening primrose flowers unfurl at dusk. After it is too dark to focus a camera, silver Y moths (*Autographa gamma*) arrive to feed on nectar from the redsepal evening primrose (*Oenothera glazioviana)* in our Surrey garden. Flash photography reveals how the moth rests on the flower, extending the long proboscis to sip nectar from the central nectar reservoir, Farnham, UK.

In 2014, a few honeybees learnt to force their way into partially open buds of this evening primrose to access virgin nectar in fading light, emerging with extensive pollen strings attached that are carried to the next flower they visit. On cooler overcast days the flowers remain open in early morning when diurnal insects visit, Farnham, Surrey, UK.

The sun bromeliad (*Fascicularia bicolor*) lives as an epiphyte, perching on trees in Chile and is pollinated by hummingbirds. The green leaves surrounding the central rosette turn brilliant red, before the pale blue flowers appear. Farnham, Surrey, UK.

When the sun bromeliad is photographed with UV flash all parts absorb UV except for the petals, which reflect it. Farnham, Surrey, UK.

CHAPTER 2

SEXY PLANTS

Flowers that open by day need to stand out from their surroundings, so they are visually attractive to their pollinators. From a distance, clusters of flowers make for a showier display than a single flower. As the pollinator approaches, the colour of the flower and scent — if present — become more important. Closer still, nectar guides direct visitors towards hidden nectar.

The flamboyant butterfly flower (*Schizanthus grahamii*) has conspicuous guidelines on the banner petal leading to the nectar source. Paso Vergara, Chile.

To ensure cross-pollination is achieved by animal vectors, flowers need to reduce the odds of pollinators homing in on them. Essentially, they need to separate from what is often a plethora of plants in surrounding vegetation. In addition to fertile flowers, some plants have sterile ones that serve to enhance the floral display to attract insects. These are seen in the outer flowers of lacecap hydrangeas and in some *Viburnum*.

Attracting pollinators

Flowers that open during the day depend on visual cues such as their shape, size, symmetry and colour. The corolla does not always provide the most attractive part of the flower, as can be seen on the dove tree (*Davidia involucrata*) which has no petals. Instead, the staminate or brush blossom flower is formed from a sphere of stamens with one perfect flower that has a style and stigma plus a ring of very small stamens. Above are a pair of unequal sized white bracts that flutter in the breeze like doves or handkerchiefs. When growing in Chinese forests amongst other deciduous trees, the distinctive white bracts are attractive to insect pollinators from a distance. Scientists in China have found that they also function as umbrellas to protect the staminate flowers from rain since when the bracts are removed the pollen is washed away. At Kew, bumblebees and honeybees both forage for pollen, but nectar is not on offer in *Davidia* flowers.

The first indication that the sun bromeliad (*Fascicularia bicolor*) is about to flower, is when the innermost-toothed leaves around the head of sessile buds, turn bright red as can be seen on page 74. After flowering, the leaves revert to green as the red fades.

Clusters of apple or pear blossom produce a more impressive display when viewed against green leaves on the tree than if they were presented as just a single flower. Even more impressive are the spectacular conical panicles of white flowers — the 'candles' — that stand erect above the new leaves of common horse chestnut (*Aesculus hippocastanum*).

Annual and perennial plants can prolong the period they remain attractive by staggering the opening of their flowers. The spider flower (*Cleome houtteana*) from South America blooms in late summer, producing fresh flowers daily that open from the bottom of the spike upwards over a period of several weeks. *Cleome spinosa* flowers open late in the day for pollination by bats. Hawk-moths also visit, but because their tongues are so long, they fail to make contact with the anthers or stigma. Orchid flowers may last for several weeks, which increases the chances of at least one visit from their infrequent pollinators.

Left
A flowering dove tree (*Davidia involucrata*) has large white bracts that attract pollinators from a distance and protect pollen from the rain. Kew Gardens, Surrey, UK.

Topside view of a spider flower (*Cleome houtteana*) shows fresh flowers open daily in a circle around the spike with the buds on top. Farnham, Surrey, UK.

Floral colour

The way we see colours is distinct from other animals. The cones in our eyes have three colour channels — red, blue and green. Unlike insects, birds and several other animal groups — including some mammals — we cannot see short ultraviolet (UV) wavelengths. Bees have three channels UV, blue and green, which enables them to see the tonal contrast of UV guides on flowers. Typically, these occur on flowers that appear as a single colour to our eyes, many on yellow flowers, including yellow evening primroses (*Oenothera* spp.).

Insect trichromatic colour evolved before angiosperm flowers appeared, so the first insects to visit these flowers had the ability to recognise floral colour.

Top
In the visible spectrum the redsepal evening primrose (*Oenothera glazioviana*) appears completely yellow.

Bottom
When lit with UV flash, the petal bases and the stigma absorb UV creating a tonal contrast with the rest of the flower, which reflects UV. Nectar in the centre fluoresces. Farnham, Surrey, UK.

Watching honeybees at work in the garden or a summer meadow, they may stick with one kind of particularly rewarding flower or they may visit different coloured flowers. Some floral visitors appear to show specific colour preferences, but these may be due to an additional feature such as scent. For example, flowers that open at dusk or during the night tend to be white or pale coloured but they also emit a strong scent and are visited largely by nocturnal moths or bats, with some beetles.

Many tropical flowers are red and are often pollinated by birds. Bees lack red receptors, so it was thought that red flowers only attract birds. However, in the Chilean Valdivian rainforest outsized native ginger bumblebees (*Bombus dahlbomii*) forage on the red flowers of Chilean lantern trees (*Crinodendron hookerianum*). Then it was discovered that these flowers, together with several other red flowering Chilean shrubs, reflect blue wavelengths, which enables bees to see the red flowers against the green foliage. In addition, as seen in photos later in this chapter, flowers with large areas of red are visited by butterflies and beetles as well as birds. *Amphicoma* beetles that visit and pollinate red bowled tulips, anemones and *Ranunculus* in the Mediterranean, have red receptors.

Scarab beetles meet and mate in a *Tulipa micheliana* flower where they have picked up pollen on their hairy bodies. Uzbekistan.

Nocturnal flowers are often white or pale-coloured with scent being the prime attractant. Green flowers exist amongst wind-pollinated grasses, whereas green is a colour rarely found in animal pollinated flowers. An exception is the emerald flower, also known as the flower of death (*Deherainia smaragdina*). Solitary green flowers appear on the evergreen shrub and emit a powerful foetid smell similar to smelly cheese, which suggests they are fly pollinated. New Zealand has a sparse insect fauna that has evolved in the absence of both honeybees (imported in 1839) and bumblebees. Half of the insect-pollinated flowers are white, with a large number of green flowers visited almost exclusively by flies.

Black flowers are comparatively rare in the wild. There is the curious bat flower (*Tacca chantrieri*) from Asia with long filaments that attract fly pollinators. A particularly striking flower is the black coral pea (*Kennedia nigricans*), an aggressive vine from Western Australia. In dull light most of the flower appears black. But when lit from behind the upright keel and wing petals appear dull red. The conspicuous lower lip has two long yellow bosses that resemble wasps — although they are visited by small honeyeaters for the nectar. These two images show freshly open untriggered flowers and flowers that have been triggered by visitors with the stamens and style sprung from the keel, as occurs with broom flowers when bees visit. Notice the yellow guides show a colour change with age as they fade from yellow to cream.

Untriggered flowers of the black coral vine (*Kennedia nigricans*). Kew Gardens, Surrey, UK.

Black coral vine (*Kennedia nigricans*) flowers triggered by a foraging honeyeater, releasing the stamens, the style and stigma. Western Australia.

Striking 9cm long red flowers of Sturt's desert pea (*Swainsona formosa*) have a conspicuous black boss. Alice Springs Desert Park, Australia.

Another leguminous flower from Australia is Sturt's desert pea (*Swainsona formosa*).This Australian desert annual produces red flowers with a black boss in response to rain. It is also bird pollinated and is the floral emblem of South Australia.

Guidelines

Once visitors arrive close to flowers with a concealed nectar source, guidelines on petals direct them to the nectar route. These are often referred to as nectar guides, which are appropriate when a flower secretes nectar, but in the case of flowers that have no nectar, they should be referred to simply as floral guides. Guidelines appear as a contrasting tone or colour in the form of spots, blotches, stripes or radiating lines. Dark blotches found on the white petal bases of the tree peony *Paeonia rockii* attract bees, while those on *Cistus* rock roses attract bees, hoverflies and beetles. Black beetle blotches in red tulips attract beetles in the Mediterranean region and in central Asia. Lines radiating in towards the nectar source are found on alstroemerias, pelargoniums, pansies and violas, while many lilies have a central nectar groove in each tepal where nectar dribbles into the base and appears blue when lit by UV flash.

Woolly rock rose (*Halimium lasianthum*) has brownish/purple patches at the base of each
petal that attract insect pollinators to the centre of the flower. Farnham, Surrey, UK.

When guides are absent, the anthers or stigmas tend to strongly absorb UV light. To our eyes, *Fremontodendron* is a completely yellow flower, but in UV light there is a distinct tonal variation. The fused petaloid sepals absorb UV, the anther filaments appear even darker in sharp contrast to the fluorescent blue nectar and small massed hairs on the anther filaments and petaloid bases which reflect UV (page 145).

Nectar spots are the most subtle type of guides appearing often as yellow or white spots. The former are thought to mimic pollen and can be seen on pickerel weed (*Pontederia cordata*). Obvious nectar guides are visible on the fall petal of each of the three reproductive units of an iris — often yellow and with added white on blue and purple irises and darker lines on yellow and pastel coloured irises. Dark patches at the base of petals of radially symmetrical open flowers can be seen in some peonies, lilies, anemones and tulips.

The blue passionflower (*Passiflora caerulea*) has downward facing anthers and stigmas that brush against the topside of larger bees as they rotate around the central column foraging on nectar. Farnham, Surrey, UK.

The corona is a striking feature of many passionflowers. When viewed from above, the filaments of blue passionflower (*Passiflora caerulea*) have distinct circular white and blue banding and a dark central zone. This serves as an attractive guide to lure insects to feed on the nectar. The versatile stamens have a very narrow attachment, which allows them to become completely flexible in the same way as spectacle nose pads, so the underside with exposed pollen makes contact with the upper surface of an insect body. Any insect that has a body that fails to contact the stamens, is an ineffective pollinator and in essence a nectar thief. Quite large bees are required to pollinate these flowers.

Colour change

We have seen that flowers need to be attractive to lure their pollinators, but they also need to signal when they are past their visit date. Some flowers have a short life and shed their petals or tubular corolla within hours of opening. Others become shrivelled up so the opening is no longer accessible. But perhaps the most interesting way in which flowers communicate with their pollinators is by a colour change taking place over the whole flower or a limited part of the flower after pollination. Floral colour change is surprisingly widespread. The production of a nectar reward costs the plant energy, so any signal that deters a visitor paying a return visit to a flower already pollinated, benefits both the flower and the pollinator.

A bumblebee homes in on a newly opened Indian horse chestnut (*Aesculus indica*) flower with a yellow patch on each of the two upper banner petals. Insects avoid older rewardless flowers with cerise patches. Kew Gardens, Surrey. UK.

Examples include pink lungwort flowers (*Pulmonaria* spp.) and *Echium wildpretii*, which open as pink and by the next day the flowers change to blue. Another colour change is for yellow flowers to fade to orange and can be seen in some evening primroses (*Oenothera* spp.), *Fremontodendron* and some legumes. Colour change may take place in a localised central area as in forget-me-nots (*Myosotis* spp.) and *Androsace*, which both have a coloured ring around the opening to the corolla tube that changes colour. Some tiny *Myosotis* flowers fade over the whole corolla.

Each flower on a Texas bluebonnet (*Lupinus texensis*) spike has a distinct white spot on the rear or banner petal. On the fifth day after a flower opens, the spots begin to turn pink before darkening to purple-red. By reducing the tonal contrast between the spot and the banner, it makes the older flowers much less attractive than the younger ones.

A particularly graphic example of colour change is seen in the pantropical invasive shrub *Lantana camara*. The first flowers to open in the flat headed umbel are the outer ones which appear yellow. These are rewarding and once a few pollen grains are transferred to a flower, this triggers the colour change to cerise by the next day, when new yellow flowers open inside the cerise ring (page 115). It does not take long for naive insect visitors to learn that the young yellow flowers are the productive ones and worth a visit. Very few visit the unrewarding cerise flowers.

Scent

Scent is a powerful attractant to pollinators especially nocturnal ones. The time of day or night when scent is released typically coincides with the time when the pollinators are active. Unlike insects that use a visual cue as they approach a flower in a straight line, insects approach a scented flower by adopting a zigzag path as they readjust their alignment to the strongest scent signal.

Flowers that emit pleasing scents to our olfactory senses include daphne, jasmine, lilies, and night scented stocks. Flowers that emit scents that can be quite noxious to us, include smells such as rotting flesh, but these are like manna from heaven for insects such as carrion flies that typically lay their eggs in rotting corpses.

Ultra Violet light signals

When visible guidelines are absent from flowers, the anthers or the stigmas are often UV absorbent. A central UV absorbent bull's-eye is widespread in radially symmetrical flowers, notably in the daisy family (Compositae) and in yellow evening primroses (*Oenothera* spp.) with the stigma lobes also absorbing UV and yellow St. John's wort (*Hypericum* spp.) with the anthers being strongly UV absorbent.

There is some doubt whether nectar that fluoresces under UV benefits floral visitors during the day; although honeybees still continue to visit almond flowers after they have lost all their petals to the wind and nectar remains. At dusk or on moonlit nights, white or cream flowers may aid animals in homing in on these flowers in dim light, although they are more likely to rely on scent, or echo location in the case of bats. Interestingly, white flowers absorb UV and so appear a dark tone all over when lit by UV.

An experiment to test the way insects react to UV absorbent patches on petals was done with a *Helianthus* flower where all the petals were removed and turned around so that instead of the dark UV areas appearing at the base towards the centre of the flower, they appeared at the outer tips. Bees responded by landing in the centre the flower and walking out towards the petal tips extending their proboscis once they reached the dark UV absorbent patches.

In visible light, a Christmas rose (*Helleborus niger*) flower appears creamy/white. When lit with UV flash, there is a distinct tonal separation, with nearly all parts absorbing UV — especially the ring of nectaries. Only the dehisced anthers reflect UV. Farnham, Surrey, UK.

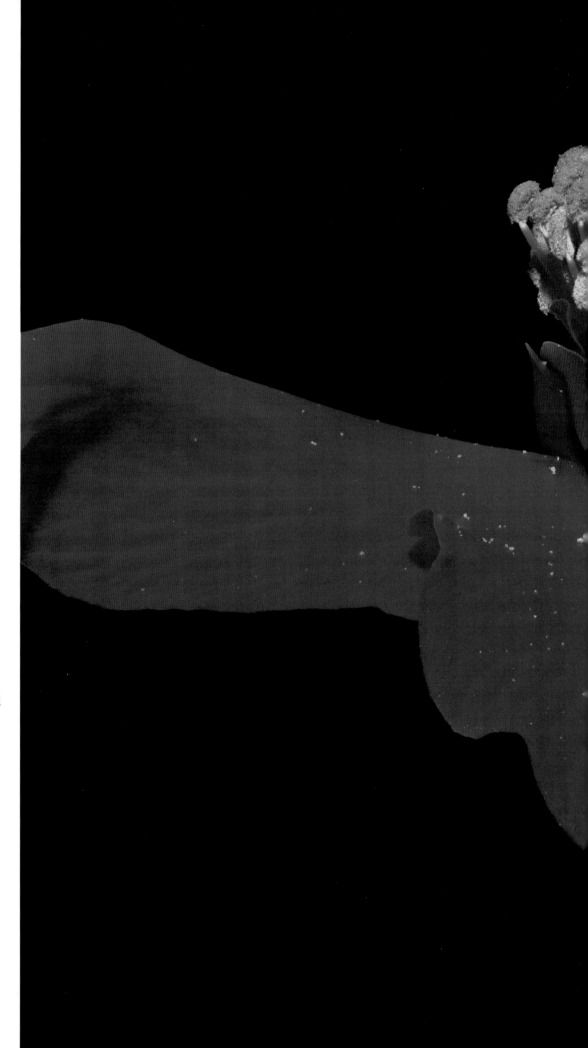

WHITE FLOWERS

Left

Magnolia grandiflora flowers control when their pollinators enter and leave by the time the petals open and close. Beetles enter when the stigmas are receptive, become trapped overnight and depart after the stamens release pollen. Kew Gardens, Surrey, UK.

Above

South African night phlox (*Zaluzianskya ovata*) flowers open fully at dusk when hawk-moths are able to access the nectar in the long corolla tube. During the day, the petal lobes close to reveal their red undersides. Farnham, Surrey, UK.

Far left
The night blooming cactus (*Echinopsis* aff. *strigosa*) flowers open at dusk. They then emit a scent that attracts nocturnal moths, which extract nectar from the long floral tube. By the morning, the flowers have wilted. Kew Gardens, Surrey, UK.

Left
A Canary Islands chiffchaff (*Phylloscopus canariensis*) feeds on nectar from white *Echium simplex* flowers in Parque del Drago, Tenerife.

YELLOW FLOWERS
Massed orange hairs form a conspicuous boss on the lower lip of common toadflax (*Linaria vulgaris*) that function as nectar guides either side of a central hairless groove. The latter guides a bee's proboscis to the nectar, stored in the spur. Selborne, Hampshire, UK.

Marsh marigold (*Caltha palustris*) flowers spotlit by early morning light. A UV light reveals a dark central bull's eye in each flower, which attracts a range of insect pollinators. WWT London Wetland Centre, Barnes, UK.

Left
Several butterflies feed from the same table
as they sip nectar from a loofah flower (*Luffa
cylindrica*), which wilts soon after mid-day.
Shunan Zhuhai National Park, Sichuan, China.

Above
The narcissus bulb fly (*Merodon equestris*) is a bumblebee mimic, here
feeding on pollen. from a woolly rock rose (*Halimium lasianthum*). The
reddish floral guides are best seen from above the flat flowers, which
drop their petals in late afternoon. Farnham, Surrey, UK.

Far left
A balloon flower (*Platycodon grandiflorus*) bud opens like an envelope flap to reveal the stamens pressed up against the style and stigma. This is an example of secondary pollen presentation. Farnham, Surrey, UK.

Left
A bumblebee (*Bombus* sp.) prises open the upper and lower lobes of a perennial sage *Salvia × sylvestris* 'Mainacht' flower to reach the nectar. Kew Gardens, Surrey, UK.

Morning glory (*Ipomoea purpurea*) flowers are blue when they open at first light and by mid-day they change to cerise before they become inrolled. Loose pollen grains have been dispersed by insects. Kew Gardens, Surrey, UK.

Gentian showing floral guide spots on outer lobes and dark spots inside the corolla directing visitors to the basal nectaries. The anthers have dehisced, but the stigma has yet to open. Farnham, Surrey, UK.

RED FLOWERS

Left

Monarch butterflies (*Danaus plexippus*) converge to feed on poinsettia (*Euphorbia pulcherrima*) in a tropical butterfly house. The red bracts are the main visual attraction and surround the tiny true flowers. Xishuangbanna, Yunnan, China.

Above

Scarab beetles (*Eulasia* sp.) are attracted to the red-bowled *Anemone bucharica* with dark stamens, where they meet and mate, picking up pollen in the process. Petal damage is caused as beetles forage — especially while awaiting a mate. Tajikistan.

Left
As a Chinese peacock butterfly (*Papilio bianor*) hovers
or perches to feed on nectar from an *Hibiscus*, the
flapping wings pick up pollen from the stamens on the
long protruding column. Yunnan, China.

Below
The red wattlebird (*Anthochaera carunculata*) is a honeyeater with a brush-
tipped tongue that mops up the nectar. Far larger than hummingbirds or
sunbirds, it still manoeuvres its body to reach the bottlebrush
(*Callistemon* sp.) reward. Quaalup Homestead, Western Australia.

FLORAL GUIDES

Left

Magnificent white trumpets of the Himalayan lily (*Cardiocrinum giganteum* var. *yunnanense*) have striking maroon guidelines, that direct visitors to the nectar source. The flowers produce viable fruits from their own pollen, transferred by pollinators. Kew Gardens, Surrey, UK.

Right

The rare clove-scented Afghanistan Juno iris (*Iris cycloglossa*) viewed from above shows the flower is made of three units. Each one has a distinct nectar guide that signals where insects need to enter for their reward. Kew Gardens, Surrey, UK.

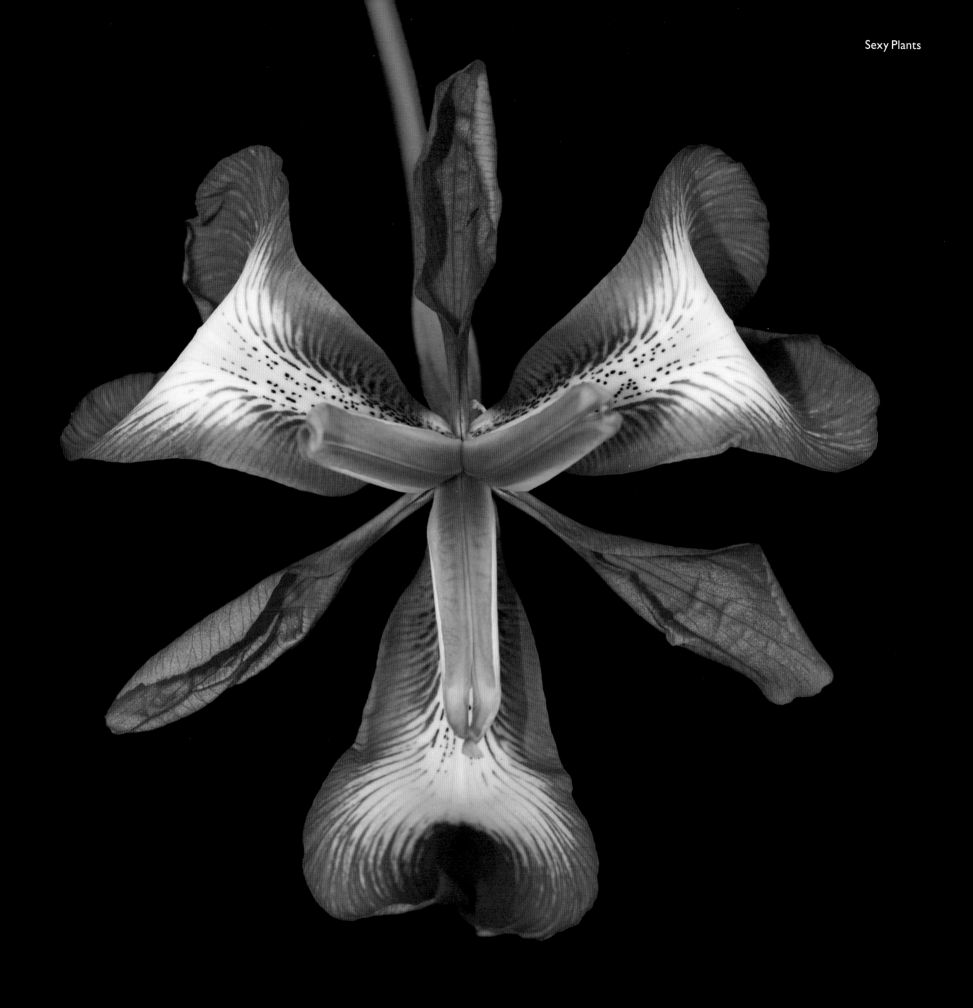

COLOUR CHANGE

Left
Tower of jewels (*Echium wildpretii*) flowers open as pink and change to blue the next day. In our garden, various bees select the rewarding pink flowers. Farnham, Surrey, UK.

Right
The yellow corona that surrounds the entrance to the blue flowers of wood forget-me-not (*Myosotis sylvatica*) fades to white with age. Kew Gardens, Surrey, UK.

Far left
When Chinese chestnut (*Xanthoceras sorbifolium*) flowers first open, a yellow patch appears in the centre, which changes colour with age via salmon to cerise. Flies and bees visit the flowers at Kew Gardens, Surrey, UK.

Left
Most red hot pokers (*Kniphofia* spp.) have orange buds that open out into yellow flowers, giving distinct coloured zones to the spike. Sunbirds visit and pollinate the flowers in South Africa. *Kniphofia uvaria* 'Nobilis' taken at Kew Gardens, Surrey, UK.

Far left
Four images show the
sequential colour change
of a red sage (*Lantana
camara*) inflorescence as it
ages. The outer florets
open as yellow, after
pollination they turn pink
the next day and inner
buds open as fresh yellow
flowers. Butterflies and
bees soon learn the yellow
ones offer the best
reward. Tanzania.

Left
Orange-petaled
passionflower (*Passiflora
aurantia* var. *samoensis*)
completely changes colour
from cream to orange-red.
It originates from New
Guinea and NE Australia
and the copious nectar is
sought by sunbirds. Kew
Gardens, Surrey, UK.

FRAGRANT FLOWERS

Left
The winter flowering shrub sweet box (*Sarcococca confusa*) emits a heady fragrance that attracts flies, hoverflies and honeybees to forage on nectar produced only by the male flowers. Honeybees, in particular, get covered with pollen as they crawl around to feed. Kew Gardens, Surrey, UK.

Right
Gold band lily (*Lilium auratum*) flowers have a conspicuous yellow nectar stripe on each tepal. This visual display, plus a strong floral scent, attracts a large swallowtail butterfly as the daytime pollinator. At night, the scent changes to attract a nocturnal hawk-moth. Kew Gardens, Surrey, UK.

SMELLY FLOWERS
The deep red lip of the orchid *Bulbophyllum echinolabium* mimics the appearance of a rotting carcass, but it is the putrid smell that lures carrion flies as pollinators. Kew Gardens, Surrey, UK.

The titan arum (*Amorphophallus titanum*) inflorescence reaches up to three metres after the tall spadix emerges from the frilly spathe. A nauseous volatile scent is wafted high into the native Sumatra rainforest where it attracts dung and carrion beetles. Taken at Kew Gardens, Surrey, UK.

VISIBLE AND ULTRA-VIOLET LIGHT

Wax flowers (*Chamelaucium* sp.) release nectar into a central reservoir surrounded by ten staminodes that alternate with the stamens. This flower originates from Western Australia.

When lit by UV flash, the petals, staminodes and stigma of a wax flower all absorb UV, while the nectar reflects it. Both taken in Farnham, Surrey, UK.

Darwin's slipper flower (*Calceolaria uniflora*) in visible light shows the white sweet lip which is eaten by the least seedsnipe (*Thinocorus rumicivorus*). As the bird bends to pluck the food body, its head brushes against the short stubby stamens.

When Darwin's slipper flower is
lit with UV flash, the entire
flower, apart from the stamens
and the fleshy lip, absorbs UV
light. Torres del Paine N. P., Chile.

Nectarless feijoa (*Acca sellowiana*) flowers provide sweet fleshy petals as a reward for the tanagers that pollinate them in Brazil. Here, a woodpigeon (*Columba palumbus*) has plucked a whole flower, complete with red stamens at Kew Gardens, Surrey, UK.

CHAPTER 3

REWARDS

Attracting a pollinator is not enough, flowers need to entice it to return. Most often, this is done by offering a reward — notably nectar and/or pollen. Other rewards include oils, resins, warm retreats, night shelters and even places to meet and mate. Some flowers are masters of deceit by mimicking real rewards. Several orchids mimic specific insects so well, the genuine ones are fooled into mating with them.

A red wattlebird (*Anthochaera carunculata*), feeds on nectar from a flame grevillea (*Grevillea excelsior)* raceme, King's Park and Botanic Garden, Perth, Australia.

By offering rewards, flowers provide an incentive to induce visitors to pay a return visit. The most widespread awards are nectar and pollen; others include food bodies, fragrances, oils and resins, nocturnal shelters and heated retreats are less widespread rewards provided by flowers.

Nectar

Even though nectar is a predominant reward, it does not occur in all flowers. Nectar is a sweet sugary solution that is the sugar source for honey and is produced by nectaries in the flower. Their position varies, but they are usually placed so the nectar-seeking visitor makes contact with the anthers and stigma as it reaches for the nectar. They may occur on the ovary, the style, the stigma, on sepals, petals or on stamens. Globules of nectar are clearly visible on the receptacle of ivy (*Hedera helix*) flowers. In flowering rush (*Butomus*) they appear as drops between the tightly packed carpels.

Nectar production coincides with the times the pollinators are most active. Visitors that feed on nectar include many kinds of bees, hoverflies, butterflies and moths, hummingbirds, sunbirds, honeyeaters, bats as well as small crepuscular and nocturnal mammals.

 Roundabout flowers, including winter aconite seen opposite, hellebores and love-in-a-mist (*Nigella* spp.) all have their nectaries arranged in a ring around the stamens. This allows insect visitors such as bees and hoverflies, to feed on successive nectaries as they perambulate around the flower and reduces the energy spent visiting many other flowers.

Nectar produced by sepal and petal nectaries may drain into a nectar spur where the nectar is stored and can be accessed legitimately only by long-tongued visitors such as butterflies, moths and hummingbirds. Here it is protected from the weather, so it is not diluted by rain nor is the consistency changed by evaporation. Since the nectar is hidden, the flower has to provide nectar guides to direct visitors to the nectar source.

Flowers with exposed nectar, do not need to provide nectar guides, since it glints in the sunlight. The downside of exposed nectar, however, is that illegitimate visitors easily thieve it and rain dilutes the nectar or even washes it away. In a dry atmosphere, evaporation results in it becoming more viscous, making it difficult to collect.

The colour of nectar varies: often it is a pale straw colour but red, brown, black, green, and blue nectar also exist. The honey bush (*Melianthus major*) has dark brown nectar and the bright red nectar of the bloody bellflower (*Nesocodon mauritianus*) appears on page 141. Why does nectar colour vary? In some cases, it may serve as a visual cue, or it may deter nectar thieves that deplete the reward without effecting pollination.

Left
A drone fly (*Eristalis tenax*) feeds on nectar from a ring of nectaries in winter
aconite (*Eranthis hyemalis*) on a sunny early spring day. Farnham, Surrey, UK.

A Nashville warbler (*Oreothlypis ruficapilla*) thieves nectar from an Indian clock vine (*Thunbergia mysorensis*) as it perches on an adjacent flowering truss; it approaches from the side and fails to contact the reproductive parts. Chiapas, Mexico.

Some nectar fluoresces in the visible spectrum when lit by ultra violet light. Honeybees continue to visit almond flowers even when all their petals are lost in windy conditions. Without petals as a visual attraction only nectar remained and when these flowers are photographed with UV flash, the nectar fluoresces blue.

Goat willow catkins (*Salix caprea*) provide a rich source of energy in early spring that attracts many insects as well as passerine birds. The trees are monoecious bearing golden male catkins or greenish female catkins. Both male and female catkins produce nectar, but only the male ones produce pollen and the female flowers produce more nectar. Insects readily pick up pollen as they forage amongst the pollen laden flowers on male catkins and bird ringers often find tits with pollen-covered faces when netting birds to ring in early spring.

On sunny days, nectar on female catkins is visible as tiny drops at the base of the many flowers on each catkin, which also fluoresces blue with UV flash. To bees and other insects able to see UV light, it may not appear this colour, but it is possible the tonal separation is enhanced between stigmas that absorb UV and the nectar. Could exposed fluorescent nectar provide a visual cue for visitors to evaluate the volume present?

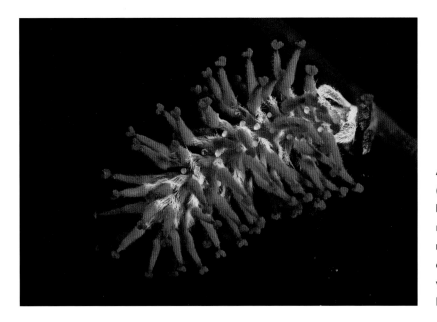

A female goat willow (*Salix caprea*) catkin lit by UV flash. Fine hairs reflect UV and the nectar flouresces blue in contrast to the stigmas which absorb UV light. Farnham, Surrey, UK.

A view inside a crown imperial fritillary (*Fritillaria imperialis*) shows the large teardrop nectaries encircled in green. Farnham, Surrey, UK.

Smaller fritillaries have linear nectaries, but the large flowers of the stately imperial *Fritillaria imperialis* have six large teardrop nectaries each encircled in green. Studies in Cambridge University Botanic Garden over three successive years, found blue tits visited and pollinated the flowers. Past sightings of tits visiting flowers always presumed they were collecting insects, not the nectar. But now there are many observations of tits feeding on nectar early in spring before the insects have begun to hatch out and multiply. After a tit lands on the broad fritillary stem it walks up to the wide opening in the bell to reach the nectar, on the way making contact with the stamens and the stigma. Also, a blackcap (*Sylvia atricapilla*) has been seen feeding on nectar from flowers growing in the wild in Iran.

Endemic bellflowers on oceanic islands are pollinated by several passerine birds, including blue tits, blackcaps and the chiffchaff. The latter pollinates the endemic Canary Island foxglove (*Isoplexis canariensis*) and the Canary Island bellflower (*Canarina canariensis*) on Tenerife.

Large individual flowers do not necessarily pull the most pollinators. Sometimes small, not particularly attractive flowers to our own eyes, may attract a range of insects. Within the space of ten minutes, four different visitors came to feed on the exposed accessible nectar from tiny yellow flowers of Jerusalem thorn in Turkey. Large flat umbels formed from massed tiny flowers, invariably attract a host of insect visitors — sometimes quite eclectic or, at times, all from the same species as when brown beetles converged on the yellow umbellifer *Ferula orientalis*, also in Turkey.

A Canary Island chiffchaff (*Phylloscopus canariensis*) stretches up to feed on a Canary Island foxglove (*Isoplexis canariensis*), Tenerife.

A fly, an ant and a ladybird visit Jerusalem thorn (*Paliurus spina-christi*) flowers for the nectar within a 10 minute period in Turkey.

The best known nectarivorous (nectar-feeding) birds are the Old World sunbirds and the American hummingbirds. In Australia, New Guinea and New Zealand, they are replaced by the honeyeaters, many of which have brush-tipped tongues that mop up nectar. Unlike hummingbirds, honeyeaters rarely hover to feed. Instead, whilst perching they may stretch across to a flower or even hang upside down. Most honeyeaters also feed on insects and some on fruit.

An increasing number of small mammals have been found to be important pollinators as they feast on nectar in South Africa and Australasia. Many are nocturnal and some crepuscular. In Australia, the pygmy possum feeds on nectar and pollen from several *Banksia* species as well as other shrubs in the Protea family. Each *Banksia* inflorescence consists of a mass of flowers. Initially they open with the looped style — not unlike the top of a paperclip. The free end with the stigma is tucked in between the anthers. As the tip of the style pulls away from the flower, it rubs against the anthers, so the pollen is transferred to the unreceptive stigma — known at this stage as the pollen presenter. As a marsupial reaches into the flowers for the nectar, their fur picks up pollen from the pollen presenters.

This hummingbird — a female white-necked jacobin (*Florisuga mellivora*) — perches on the long spathe-like bract to feed on an expanded lobsterclaw (*Heliconia latispatha*), Bosque del Cabo, Costa Rica.

Far left
A Tasmanian pygmy possum (*Cercartetus lepidus*) extends a narrow tongue to feed on silver banksia (*Banksia marginata*) nectar at night. This is one of several marsupials, together with honeyeaters, that pollinate silver banksia. S E Australia. © C. Andrew Henley.

Left
The wide mouth to the bell of the cup and saucer flower (*Cobaea scandens*) allows a bat to reach the nectar at the base of the bell. Bat flowers tend to have the stamens and stigma protruding from the bell, so the bat brushes against them on entry. Farnham, Surrey, UK.

Pollen

Pollen-rich flowers such as peonies attract a succession of insects, which converge to feed directly on the pollen or to collect it for brood food. In addition to pollen feeders and collectors that aid pollination, pollen is susceptible to loss by wind, rain, and illegitimate visitors that play no part in pollination. The colour of pollen varies. Typically, it is yellow or orange, coloured by flavonoid pigments that provide protection against UV. However, if you watch honeybees or bumblebees collecting pollen and look at their pollen baskets, more than one colour can be seen during a fine weather period in summer. Other colours of pollen include white, grey, blue, pink, red and brown. Beekeepers use this as a way of monitoring on which plants the bees are feeding.

Honeybee (*Apis mellifera*) with an overflowing pollen basket forages on pollen in peony, (*Paeonia officinalis* subsp. *microcarpa*) picking up pollen all over its body. Pollen has also been transferred to the pink stigmas. Kew Gardens, Surrey, UK.

It is easy to see when bees are working with pollen and not feeding on nectar because their tongue is not extended and they do not forage deep inside the flower. Instead, they either crawl amongst the anthers or walk on top of them to forage. The front legs are used to pick up pollen, which is transferred to the corbicula or pollen basket on each hind leg of honeybees, bumblebees, stingless bees, and orchid bees. In flowers with a heavy pollen load, a honeybee periodically lifts off from the flower to hover, so it has plenty of free space to transfer pollen to an overloaded pollen basket.

How do visitors gauge whether it is worth visiting a flower or not? Exposed pollen and nectar becomes visible at close range. Rain soaked pollen needs to dry out before it is transferred to a visitor, so this is why many flowers respond to rain by closing their petals or by having their anthers tucked away beneath a protective overhanging hood. This means that both the route to concealed nectar beneath flaps or in spurs, as well as anthers that lie hidden has to be signalled — like signposts on a road — by floral guides.

The foxglove (*Digitalis purpurea*) has dark red spatter spots encircled in white on the lower lip that are continued towards the back of the flower that serve as floral guides. When the sun shines through the flowers, it illuminates the tunnel inside — highlighting the scattered spots.

A garden bumblebee (*Bombus hortorum*) alights on a foxglove (*Digitalis purpurea*) flower to forage. The downward pointing flowers with the large upper lip, protect the reproductive parts from rain. Kew Gardens, Surrey, UK.

Nocturnal shelters

Not until I visited Turkey some years ago did I appreciate how insects use flowers as nocturnal shelters. Both male solitary bees and scarab beetles use flowers in this way in the Mediterranean region. If they simply group together on dead flower heads such as a *Primula* or in between individual flowers in a cluster, the benefit is clearly one-sided to the insects. However, when they enter an intact flower, they can pick up pollen. This method of pollination is now known to occur in some irises by male eucerine bees at dusk.

A male eucerine bee turns to leave an *Iris iberica* subsp. *elegantissima* flower at dusk; note pollen on the bee's thorax and wing, Nakhchivan. © Diana Barrie.

Originating from the Middle East, Oncocyclus irises have large self-incompatible flowers with hidden pollen and no nectar. Solitary male bees seek large dark (red-brown, purple and black) iris flowers at dusk as night-time roosts. Male bees (mostly *Synhalonia spectabilis*) enter several *Iris atropurpurea* pollination tunnels before sunset. Why they check out several flowers before settling down for the night is puzzling, since they do not reject occupied flowers with up to twelve bees found in a single iris. When bees exit the flower, pollen is visible on their topside, so pollination probably takes place as the bees switch flowers before roosting. *Iris atropurpurea* absorbs light across the UV spectrum, so it is presumed bees see the flowers as 'bee-black' and select the dark shelter mimics that separate tonally from adjacent vegetation. Male bees emerge to coincide with the time the irises are in flower.

Other more sophisticated shelters involve the production of a strong attractive scent that lures beetles to flowers at the precise time when the stigmas are receptive and trap the beetle pollinators by petal closure, before their release the next day. This occurs with *Magnolia grandiflora*, the giant water lily (*Victoria* spp.) and lotus lily flowers (*Nelumbo nucifera*). The latter two flowers also generate heat that keeps the beetles active so they pick up pollen as they move around inside the flower.

Deceitful flowers

Some flowers attract their pollinators by structurally mimicking real rewards or sometimes emitting a deceptive scent. Only male flowers offer pollen as a reward, therefore pollen-collecting bees should favour male flowers, which would mean this would reduce the opportunities for cross-pollination. But many bee-pollinated flowers have developed yellow spots that mimic pollen in both their colour and their UV absorption. Experiments using artificial flowers and naive bumblebees (*Bombus terrestris*) found the bees were attracted to pollen-yellow floral guide spots positioned near the base of the petals.

Pickerel weed shows the pair of yellow anther mimic spots on the banner petal. Farnham, Surrey, UK.

An example of a flower with yellow anther mimic spots is the aquatic pickerel weed (*Pontederia cordata*), which has a pair of yellow spots on the blue banner petal. In fact, the spots are much more conspicuous to our eyes than the actual anthers, which are the same colour as the petals.

The flowers of African hemp (*Sparmannia africana*) have sensitive stamens that move in response to touch. When the white petals open, the stamens are bunched up together, but if triggered by an insect or any object, they separate. Encircling the true fertile stamens is a whorl of yellow staminodes with conspicuous nodes that resemble a string of yellow irregular beads. By mimicking 'feeding anthers' they infer the pollen load is more abundant and rewarding than in reality. This stimulates bees to begin pollen-collecting movements, repeatedly touching the stamens and stigma with the underside of their abdomen.

Triggered stamens of African hemp (*Sparmannia africana*) show outer staminodes that mimic feeding anthers. Kew Gardens, Surrey, UK.

Begonias have separate male and female flowers that produce no nectar. The male staminate flowers attract pollinators with their yellow anthers and yellow pollen. The female flowers produce yellow stigmatic lobes that mimic the anthers in their shape and colour, thereby fooling pollinators into visiting their rewardless flowers.

Anthers are not the only floral parts that are mimicked by flowers. Many pelargoniums have obvious nectar guides in the form of dark lines on the paler banner petals, but *Pelargonium tricolor* has dispensed with these. Instead, it has a blackish warty patch at the base of each red banner petal, that functions as a false nectary, reflecting spots of light that mimic highlights in real nectar. This attracts the bombyliid fly *Megapalpus capensis* as a pollinator.

An even more impressive example of nectary mimicry is seen in the grass-of-Parnassus or bog-star (*Parnassia palustris*). Lying above the white petals, are five staminodes that form golden pseudonectaries, tipped with swellings that glisten in the sunlight. Although completely dry, they nonetheless attract flies.

Pelargonium tricolor has a blackish warty patch at the base of each red banner petal, that acts as a false nectary by reflecting highlights that mimic those in real nectar. Kew Gardens, Surrey, UK.

The masters of deceit, however, are the *Ophrys* orchids that mimic insects in their shape and colour. The Mediterranean mirror orchid (*Ophrys speculum*) is pollinated exclusively by males of the wasp *Campsoscolia ciliata*. A reddish-brown hairy lip rim surrounds a blue shiny patch akin to the sky reflection on the wasp's wings. This — plus a cocktail of floral scents that mimic the mating pheromones produced by the female wasp — tricks a male wasp into believing it has found a potential mate. So, the male wasp attempts to mate with a flower and since it proves unsuccessful, it moves on to other mirror orchids, thereby pollinating them in the process, without gaining any reward.

A grass-of-Parnassus flower shows golden pseudonectaries that offer no reward, but are still attractive to flies. Gares, Dolomites, Italy.

NECTAR

Left
When the Indian clock vine (*Thunbergia mysorensis*) grows up a pergola or over an arbour, new flowers open daily on long racemes, producing nectar that overflows from the yellow throats attracting sunbirds in Asia. Kew Gardens, Surrey, UK.

Right
Flowering rush (*Butomus umbellatus*) with pollen-laden anthers surround the tightly packed pink carpels with white stigmas. Six clear nectar globules are also visible. Kew Gardens, Surrey, UK.

A Cape white-eye (*Zosterops pallidus*) sips
nectar from honey bush (*Melianthus major*),
flowers, which contain dark nectar.
Kirstenbosch NBG, South Africa.

The bloody bellflower (*Nesocodon mauritianus*) produces rare red nectar. Only a few plants remain on steep sided cliffs in Mauritius and Kew is helping to propagate them. The pollinator is unknown, but the native ornate day gecko visits another plant also with red nectar.

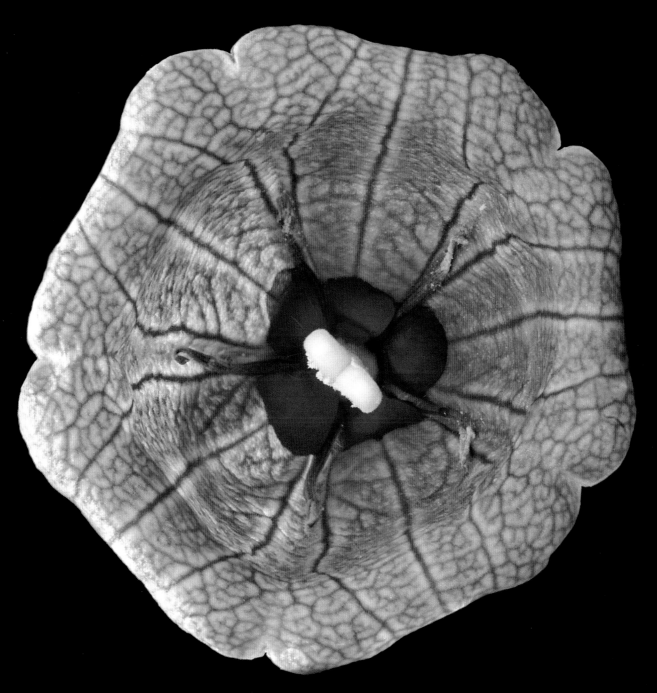

A backlit Congo cockatoo flower (*Impatiens niamniamensis*) reveals the nectar level in the spur. Native to central Africa, when a sunbird probes for nectar, pollen is deposited on its head and carried to the stigma of an older flower. Taken in Kew Gardens, Surrey, UK.

Late in the day, a banded mosquito (*Culiseta annulata*) feeds on nectar from a green-flowered helleborine (*Epipactis phyllanthes*), but does not pollinate the orchid. It self-pollinates when fragments of the friable pollen sacs fall onto the stigma. Alice Holt Forest, Surrey, UK.

Yellow flowers of the California flannel bush (*Fremontodendron californicum*) are formed from five fused petaloid sepals, each with a nectary at the base, and are pollinated by native solitary bees. This cultivar, *Fremontodendron* 'California Glory', was photographed in Farnham, Surrey, UK.

NECTAR IN UV

When *Fremontodendron* 'California Glory' is lit with UV flash, the position of the nectaries and overflowing nectar is revealed as it fluoresces blue in UV against the rest of the flower which absorbs UV. Farnham, Surrey, UK.

Yellow nectar, barely visible in normal
light, dribbles from the base of the
stamen filaments of the Arabian
starflower (*Ornithogalum arabicum*).
Farnham, Surrey, UK.

When lit with UV flash, the extent of the nectar in the Arabian starflower (*Ornithogalum arabicum*) is immediately apparent on two flowers since it fluoresces yellow in contrast to the flower, which absorbs UV. Farnham, Surrey, UK.

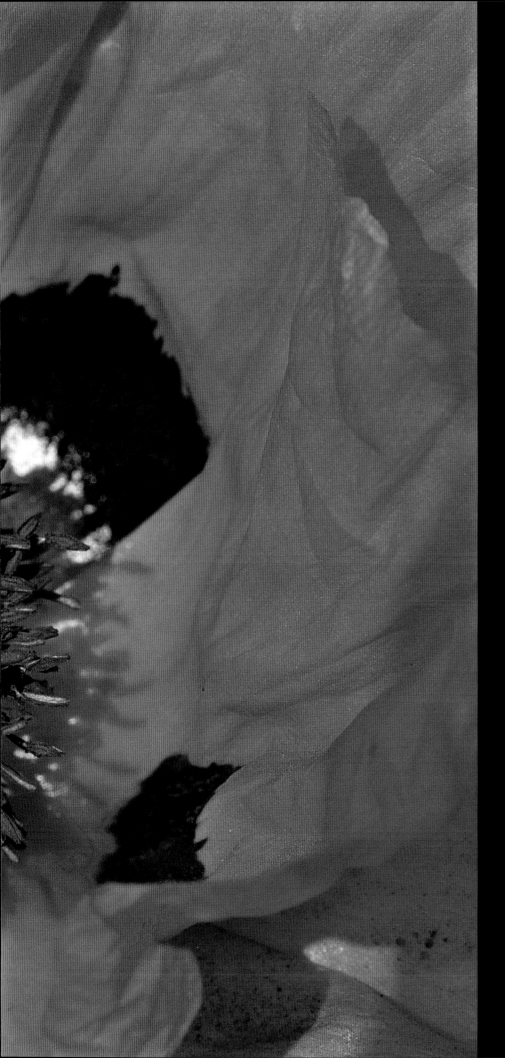

POLLEN

A buff-tailed bumblebee (*Bombus terrestris*) forages on an oriental poppy (*Papaver orientale*) with a dark plum coloured pollen load, Kew Gardens, Surrey, UK.

A newly opened flower of the New Zealand tree fuchsia (*Fuchsia excorticata*) has anthers that carry rare bright blue pollen and hang down above the bright yellow stigma. Wakehurst Place, Sussex, UK.

Right
Pollen feeders on Californian tree poppy (*Romneya coulteri*) are a male dead head hoverfly (*Myathropa florea*) and a female leaf-cutter bee (*Megachile centuncularis*). As the latter sits on the stigmatic disc, pollen is transferred from special pollen-carrying hairs beneath the abdomen. Kew Gardens, Surrey, UK.

Left
A honeybee and a bumblebee both forage for nectar on the same love-in-a-mist (*Nigella damascena*) flower, with the honeybee picking up pollen as it is swiped by the downward facing anthers. Kew Gardens, Surrey, UK.

Above
When lit with UV flash, the stamens and stalk of a love-in-a- mist (*Nigella damascena*) flower reflect UV and contrast dramatically with the rest of the flower, which absorbs UV. Kew Gardens, Surrey, UK.

BUZZ POLLINATION

Left

A buff-tailed bumblebee (*Bombus terrestris*) hangs upside down from a strawberry tree (*Arbutus unedo*) flower buzz pollinating it with short audible buzzes, to release the pollen that falls onto its body. Kew Gardens, Surrey, UK.

Above

Before the leaves open, large showy flowers of the buttercup tree (*Cochlospermum vitifolium*) appear, each with 150–200 orange stamens. Pollen is the sole reward, gleaned by bees as they vibrate the stamens to release it by buzz pollination. Chiapas, Mexico.

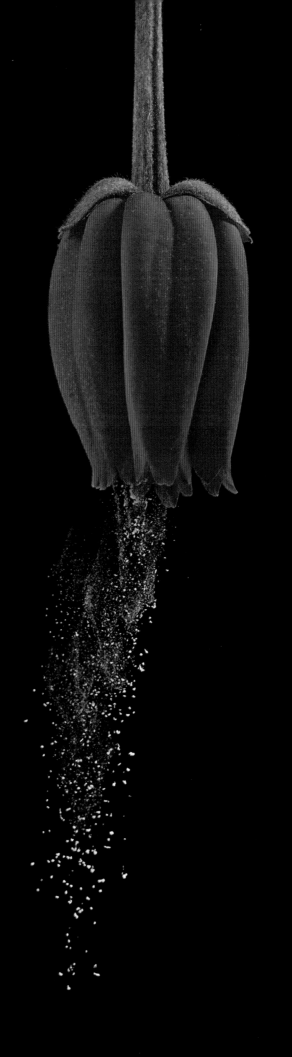

High-speed flash freezes the pollen
release from the anther pores of a Chilean
bellflower (*Crinodendron hookerianum*)
triggered via mechanical vibration, which
replicates buzz pollination or sonication
by a bumblebee. Farnham, Surrey, UK.

Mechanical vibration of a snowdrop
(*Galanthus nivalis*) triggers pollen release from
the stamens that open at their apex. Taken
with high-speed flash, Farnham, Surrey, UK.

NIGHT SHELTERS
Two long-horned bees crawl into a *Gladiolus atroviolaceus* flower for a place in which to shelter for their nocturnal roost in Turkey.

A sleeping assemblage of scarab beetles congregate on a *Primula* head with faded flowers, late in the day in Turkey.

Above
Large bellflower bees (*Melitta haemorrhoidalis*) visit campanulas to feed during the day and here two male bees are sleeping at night inside a peach-leaved bellflower (*Campanula persicifolia*). Farnham, Surrey, UK.

Right
Green lacewings use the fragrant cup and saucer flowers of *Magnolia* 'Phelan Bright' as a night shelter. Kew Gardens, Surrey, UK.

WARM RETREATS

Left

The scented white flower of the Santa Cruz waterlily (*Victoria cruziana*) opens in the evening, attracting beetle pollinators that become trapped inside when the flower closes. The floral prison generates heat; changes colour to pink and sheds pollen before it releases the beetles with their pollen loads. Kew Gardens, Surrey, UK.

Above

A newly opened lotus lily (*Nelumbo nucifera*) has receptive stigmas before the stamens elongate and shed pollen. Beetle pollinators become trapped inside and are kept warm by the flower heating up and maintaining a temperature of 34–36°C. Paya Indah Wetlands, Malaysia.

MEETING PLACES

A male wool-carder bee (*Anthidium manicatum*) attempts to mate with a female whilst she visits a rusty foxglove (*Digitalis ferruginea*), Kew Gardens, Surrey, UK.

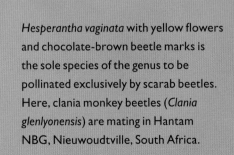

Hesperantha vaginata with yellow flowers and chocolate-brown beetle marks is the sole species of the genus to be pollinated exclusively by scarab beetles. Here, clania monkey beetles (*Clania glenlyonensis*) are mating in Hantam NBG, Nieuwoudtville, South Africa.

CHAPTER 4

ANTI-POLLINATORS

Warm, wind-free days are perfect for insect pollinators; whereas cold days, rain and wind curtail their activity. An assortment of animals influences plant-pollinator interactions: most obviously, the animals that pluck flowers to eat or that prey upon pollinators. There are also nectar robbers that thieve nectar without pollinating the flower, while some animals even affect the behaviour of true pollinators.

A spotted blister beetle (*Ceroctis capensis*) feeds in the rain on *Gladiolus alatus* causing damage to the petals. South Africa.

Rain fell continuously as we drove past many wild golden spider lilies (*Lycoris aurea*) in Yunnan, China, so no visitors were spotted. In Japan, swallowtail butterflies visit them.

Once rain falls, insects cease foraging to take shelter, whereas birds are able to continue foraging providing the rain is not too heavy. Frost also damages early spring flowers such as camellias making them unattractive and even causing damage to floral parts of open flowers and bud drop.

Floral predators

In tropical rainforests beetles are the most abundant type of pollinator and a visit to South Africa, Turkey or Central Asia, reveals beetle visitors in many a flower. Indeed this is where they often meet and mate. Although beetles are proven pollinators, they also feed on the petals. A few nibbles along the petal edges does not diminish from the visual appealing of the flower to other insects; but several beetles can cause enough damage, for the petals to become flaccid and shrivel making them less appealing to other pollinators.

Some larger bees, in particular, take a shortcut to reach the nectar by piercing the base of the corolla tube. Unless more nectar is secreted, this depletes the reward for legitimate pollinators, which enter through the front opening. Honeybees will sometimes utilise existing robbers' holes as a speedy access to a nectar source.

Both parakeets and parrots feed on nectar from flowering trees in the gallery forest in the South Pantanal, Brazil. Crimson-bellied parakeet (*Pyrrhura perlata*) flocks converge to feed on the nectar from the pink trumpet tree (*Tabebuia heterophylla*) flowers that appear on the leafless tree. The birds either pierce the base of the flower or pluck them from the tree and squeeze the bottom to access the nectar. Toco toucans (*Ramphastos toco*) extract nectar from these flowers by holding them at the base and squeezing them with their outsized, but highly dextrous bill. They also eat the flowers as well.

Agile monkeys that move through tree canopies soon learn the whereabouts of rewarding trees in a particular season — whether it be to visit flowers for their nectar or to consume the flowers themselves. In Madagascar, lemurs too seek nectar during the dry season. In Europe, rabbits, deer and squirrels eat wildflowers early in spring when the choice of food is limited. They also invade gardens to feast on flowers, with roe deer munching their way through tulip and geranium flowers and rabbits are known to eat snowdrops.

Pollinator predators

Some predators have learnt to play a waiting game to gain a free lunch by lurking inside or beneath the flower until a pollinator lands within their grasp. Praying mantids are effective predators that blend in with vegetation and adopt a statuesque attitude. Mantids have the ability to suddenly pounce on their prey by extending their lengthy front pair of legs, which end in clasps. Their sharp mandibles then make mincemeat of their prey.

A well camouflaged white crab spider (*Misumena vatia*) has captured a fly visiting white *Pieris formosa* var. *forrestii* flowers. Farnham, Surrey, UK.

A white crab spider (*Misumena vatia*) preys upon a green hairstreak butterfly (*Callophrys rubi*) on rosy garlic (*Allium roseum*) Picos de Europa, Spain.

Crab spiders, although smaller, are capable of capturing prey much larger than themselves — including butterflies. These spiders often have a cryptic body colour that matches the flower they use to ambush their prey, typically white, yellow or pink. This helps them to blend into their background. Is this more advantageous to them in fooling their prey, or to ensure they pass unnoticed by *their* predators? Crab spiders simply sit atop a flower or underneath it with their first two pairs of elongated legs held out ready to clasp a hapless visitor within reach. Once caught, fangs inject venom into the prey to immobilise it, so the body fluids can be sucked out and the empty corpse discarded. As well as butterflies, crab spiders prey upon flies, hoverflies and bees.

In our Surrey garden, crab spiders are very adept at catching hoverflies that linger to feed on lily pollen; the evidence remains as corpses scattered around the potted lilies the following morning. I have also seen crab spiders at work on marsh helleborines (*Epipactis palustris*) in a year when red soldier beetles were the chief visitors and pollinators — seen to fly off with pollinia attached to their wing cases. However, since crab spiders are neither abundant nor highly mobile, they are unlikely to play a major part in influencing plant-pollinator interactions.

The more numerous and aggressive weaver ants (*Oecophylla smaragdina*), on the other hand, can cause a decline or an increase in pollinator visits. When weaver ants construct their nest in plants of the tropical fruit rambutan (*Nephelium lappaceum*), pollinators avoid these flowers. Yet, large carpenter bees (*Xylocopa* sp.) which pollinate the pink flowers of the shrub *Melastoma malabathricum*, are too large to be attacked by the ants, which deter small bees that remove pollen but are inefficient pollinators.

Many bees are lost to attractive bee-eaters that manoeuvre with great agility to catch these insects in mid-air. Beewolves are solitary wasps that live up to their name, although their prey is used as brood food for their larvae since the adults feed on nectar. The European beewolf (*Philanthus triangulum*) seeks out honeybees, which it catches either whilst they forage in flowers or in mid-air. The beewolf injects venom into the honeybee to paralyse it, before it flies off clasping the bee beneath its body. Up to six honeybees are stored in one brood tunnel where a single egg is laid.

An interactive food web

The showy pink flowers of the mother of cocoa tree (*Gliricidia sepium*), so called because it is used as a shade tree for cocoa, caught my eye during a February visit to Chiapas in Mexico. The most numerous visitors were large carpenter bees, which were busily feeding and pollinating the pink flowers, each with a broad yellow stripe on the upper banner petals. Hummingbirds darted in and out to feed on the nectar, never lingering for long on any flower cluster. The occasional attractive Altamira oriole also visited for the nectar. On the second afternoon a boat-billed flycatcher arrived pausing on a branch before it plucked a feasting carpenter bee. The bird repeatedly bashed the bee against a branch to stun the insect before it could reposition the bee in the bill, so it could be swallowed headfirst. Three more bees met the same fate before the flycatcher departed.

Next, a small group of Aztec parakeets came to plunder the flowers by plucking them individually and eating them. Only then did I appreciate just how many flowers had been removed by these birds, probably over a period of several days; but since a single tree can produce up to 30,000 flowers, it can afford to lose a few. Here is an example of how different animals converge on a flowering tree to form an interactive web.

Cornfields once harboured colourful wildflowers as this mix of corn poppies, cornflowers and ox-eye daisies planted in a Surrey garden. The widespread use of herbicides means they are now confined to field margins or hand sown patches on open ground.

NECTAR ROBBING
A female carpenter bee (*Xylocopa caffra*) bee inserts its proboscis into the base of an African foxglove (*Ceratotheca triloba*) flower to rob the nectar, without pollinating the flower. Kirstenbosch NBG, South Africa.

A male carpenter bee (*Xylocopa caffra*) grasps a ninepin heath (*Erica mammosa*) flower to rob nectar by using the proboscis to pierce the corolla. This limits the reward for pollinators, which access via the open flower. Harold Porter BG, South Africa.

POLLINATOR PREDATORS
The solitary wasp known as the European beewolf (*Philanthus triangulum*) preys upon honeybees, captured for brood fodder. Here a female beewolf rests after capturing a honeybee in midflight. Farnham, Surrey, UK.

A white crab spider (*Misumena vatia*) rests on an *Enkianthus campanulatus* flower poised with the first two pairs of legs outstretched ready to ambush an insect visitor. Farnham, Surrey, UK.

POLLINATOR

Left

A carpenter bee feeds legitimately on nectar via the front of a mother of cocoa (*Gliricidia sepium*) tree, which is used as a shade tree for cocoa. The banner petal on each flower has a distinctive yellow stripe that serves as a nectar guide for this pollinator. Chiapas, Mexico.

PREDATOR

Above

One of several carpenter bees (*Xylocopa fimbriata*) feeding on a mother of cocoa tree caught by a boat-billed flycatcher (*Megarynchus pitangua*). Before being swallowed, the bee was bashed repeatedly on a branch. Chiapas, Mexico.

VISITOR
An Altamira oriole (*Icterus gularis*) feeds legitimately on nectar from a mother of cocoa tree flower. Chiapas, Mexico.

FLOWER FEEDERS

A small flock of Aztec parakeets (*Eupsittula astec*) flew into a mother of cocoa tree to feed on the flowers by plucking them off the tree. Chiapas, Mexico.

In the middle of the dry season in South Pantanal,
Brazil, pink trumpet tree (*Tabebuia heterophylla*)
flowers are eaten by toco toucans (*Ramphastos
toco*), which also suck nectar from the base.

A colour co-ordinated caterpillar feeds on a
Cape buttercup flower (*Sparaxis elegans*), which
has coiled anthers that twist around the style.
Nieuwoudtville, South Africa.

Left
A hanuman langur (*Semnopithecus entellus*)
sits on a branch of a mahua tree (*Madhuca
longifolia*) to feed on flowers it has plucked,
thereby preventing pollination and reducing
seed set, Khana National Park, India.

Above
A ring-tailed lemur (*Lemur
catta*) feeds on a cactus flower
after plucking it from the
plant, Berenty, Madagascar,

Scarab beetles settle on an *Iris vicaria* flower to roost late in the day. The damage to the flower, caused by beetles feeding on it, make it unattractive to potential pollinators. Tajikistan.

POLLINATOR

A pollen encrusted bee exits from an attractive intact *Iris*

CHAPTER 5

FLOWERS — THE FUTURE?

Throughout the world, natural plant communities are lost to agriculture and urbanisation or changed by grazing stock and invasive aliens. Critically endangered plants are being saved from extinction by propagation and repatriation. Gardens can provide valuable nectar and pollen resources — although many cultivars bred to attract gardeners, offer nothing for pollinators.

Each tiny lily pad of the pygmy Rwandan waterlily (*Nymphaea thermarum*) reaches one centimetre (0.4 in) in diameter and the flowers close by mid-day. Kew Gardens, Surrey, UK.

Flowers — the future?

St Helena ebony (*Trochetiopsis ebenus*) with secondary
pollen presentation on the petal rims of a newly opened
flower. Inside are five carmine staminodes and a central
cream stigma. Kew Gardens, Surrey, UK.

Special endemic island flora confined to a limited habitat is more susceptible to the risk of becoming critically endangered from introduced alien species outcompeting them and herbivores that destroy them. Critically endangered species — some from just one or two plants remaining in the wild — have been saved from extinction by propagation and repatriation into the wild wherever possible. Alternatively an *ex situ* stock is maintained.

Back from the brink

The pygmy Rwandan waterlily (*Nymphaea thermarum*) — the smallest waterlily in the world — grew in damp mud fed by hot spring water. After the water was exploited for agriculture and the warm mud dried up, the species became extinct in the wild. Before then, a few plants and seeds were collected by Bonn Botanical Garden in Germany. Seeds planted in pots submerged underwater germinated, but produced feeble seedlings. When a horticulturalist at Kew replicated the natural conditions, he successfully propagated the first plants. They flowered in November 2009 in pots placed inside a larger container filled with warm water so it overlapped the compost and the leaves spread out over the mud, as in Rwanda. The flowers open in the morning and close at mid-day suggesting the natural pollinator may be a diurnal insect. The plants at Kew can be self-fertile, but they are also hand pollinated.

Trochetiopsis is an endemic genus on St Helena — an island in the South Atlantic formed from mid-Atlantic Ridge eruptions, now an eroded summit of a composite volcano. Three species of these evergreen woody plants once existed — the blackwood ebony (*T. melanoxylon*) extinct since the end of the 18th century, the redwood (*T. erythroxylon*) now extinct in the wild and the St Helena ebony (*T. ebenus*) is Critically Endangered. Self-pollination of the redwood led to a high level of inbreeding, resulting in shrubs now reaching a third of their normal height.

Instead of the pollen being presented directly from the anthers, it is transferred to the petal margins in the bud stage. This is an example of secondary pollen presentation. It is not known whether the native pollinators still exist or are now extinct. Kew maintains *ex situ* populations of both the redwood (*T. erythroxylon*) and the St Helena ebony (*T. ebenus*). Additionally, two former Kew horticulture students manage populations repatriated to conservation areas on St. Helena.

Café marron (*Ramosmania rodriguesii*) is a shrub endemic to the Indian Ocean island of Rodrigues — part of the Mascarene Islands. Using cuttings taken from the single plant that remains on the island, Kew has successfully propagated the plant. The self-incompatible flowers prevented the lone plant being fertilised with its own pollen, until a horticulturalist at Kew discovered a way round this problem, resulting in a small number of viable seeds. These produced seedlings, amongst which cross-pollination was possible, thereby gaining a higher seed yield, some of which was returned to Rodrigues for propagation there.

Café marron (*Ramosmania rodriguesii*) from Rodrigues, can be seen at Kew, where it has been successfully propagated. Kew Gardens, Surrey, UK.

Invasive aliens

Plants in their native countries are kept in check by herbivores and diseases but some plants when introduced — both accidentally and intentionally — to other countries, colonise at a rapid rate without any natural means of controlling them. Referred to as non-native invasive species (NIS), eradication of these plants, if possible, is costly. NIS affect food chains as well as plant-pollinator interactions.

A large ginger bumblebee (*Bombus dahlbomii*) is the sole *Bombus* species native to southern South America, where it pollinates many native plants. The buff-tailed bumblebee (*Bombus terrestris*), one of the most numerous bumblebee species in Europe, is bred to pollinate greenhouse crops, and has been imported to many countries and areas where it is not native. It was introduced to Chile in 1998 and has since entered Argentina. Now feral, it is spreading at a rate of 275km per year. Where it occurs, *Bombus dahlbomii* disappears within weeks. The sudden death of the native bee is thought to be due to a parasite carried by the buff-tailed bumblebees, that has no adverse effect on them. In 2008, the Australian government banned the importation of live buff-tailed bumblebees to Australia fearing it would present a significant threat to native fauna and flora should it become a feral species.

A giant ginger bumblebee (*Bombus dahlbomii*) forages on a lantern tree (*Crinodendron hookerianum*) flower. Parque Nacional Alerce Andino, Chile.

Red sage (*Lantana camara*) is a NIS pantropical shrub originating from South America that has spread throughout tropical and subtropical countries. The colourful flower heads attract a host of insect pollinators as well as sunbirds and hummingbirds. All visitors soon learn the young yellow flowers offer a nectar reward, unlike those that change colour to deep pink after pollination.

Centuries after whalers and sealers released rabbits and goats on remote oceanic islands in the southern hemisphere as food for ship-wrecked mariners, their impact on endemic plants has been catastrophic. In addition, rats sometimes jumped ship and invaded land as well. Goats are particularly destructive because not only do they munch their way through herbaceous plants as well as shrubs, but can cause serious erosion if they destroy the ground cover.

After *Hibiscus* plants were isolated on the oceanic islands of Hawaii, Madagascar and Mauritius for long periods, new endemic species evolved. Many are now extinct or highly endangered. On Hawaii, populations crashed as a result of volcanic eruptions and browsing by introduced goats.

Perfect flowers for pollinators

Neither natural habitats nor gardens remain frozen in time; they both evolve as more vigorous species invade and compete with smaller slow growing ones. As urbanisation increases, gardens and parks in urban areas become increasingly important as reservoirs of plant and pollinator associations. It helps to plant pollinator friendly plants (a good garden centre should provide sound advice) that provide nectar and pollen sources not just in spring and summer, but throughout the year. Avoid highly bred cultivars, as selective breeding — especially of double flowers — have made some varieties of little or no use to insects because either they have lost their reproductive parts or the petals are so tightly packed the insects cannot access them. Planting only cultivars with double flowers that offer no rewards will look more attractive than a paved front yard, but as far as pollinators go, it is equally unrewarding.

These do not need to be restricted to native plants. Non-native flowers such as the strawberry tree (*Arbutus unedo*) and various mahonias provide valuable winter food resources after native shrubs have finished flowering. This enables *Bombus terrestris* to produce a late season second generation in southern Britain. If *Arbutus* flowers persist until late December, bumblebees will continue to forage on sunny windless days.

Additionally, over 6,000 British churchyards are now managed as sacred eco-systems. By mowing once or twice a year and no longer using herbicides or pesticides, neatly manicured lawns have been replaced with wildflower displays: snowdrops, crocuses and wild daffodils in early spring, followed by an array of summer meadow flowers. Most old churchyards were originally meadowland untouched by ploughing. With old, relatively undisturbed grasslands now rare in the UK, these living churchyards have become valuable wildlife oases, providing rewards for invertebrate pollinators, which in turn provide food for vertebrates such as amphibians, reptiles and bats.

In early spring, honeybees and bumblebees busily forage on crocuses that carpet the ground amongst gravestones in the Bourne Old Churchyard, Farnham, Surrey, UK.

The future

Opened in 2000, Kew's Millenium Seed Bank situated at Wakehurst Place in Sussex, has saved seeds from over 13% of worldwide plant species. By 2020, the aim is to increase this to 25% stored in the large underground vault. Plant species are lost when primary vegetation is cleared for monoculture crops or urbanisation and because of climate change. Yet plants are not only essential for food, they also provide medicines, clothing as well as building materials for housing and transportation.

In addition, green plants play an important role in helping to combat climate change. During the process of photosynthesis, energy from sunlight, absorbed by the green pigment chlorophyll, aids the production of nutrient sugars from carbon dioxide and water. Oxygen is released back into the atmosphere, thereby enhancing the oxygen levels.

Pollinators have been dwindling for the last two decades. The worldwide decline in bees is partly due to excessive use of insecticides — notably neonectinoids — and to urbanisation. Other contributory factors are the parasitic varoa mites causing colony collapse disorder (CCD) in honeybee hives and extreme weather conditions. Habitat fragmentation means that pollinators have to fly greater distances to feed or collect their reward, which reduces the number of plants that are visited. The rapid spread of alien species can compete for growing space and for pollinators.

After excessive use of pesticides in the Chinese province Sichuan, apple orchards in a 60 km valley, failed to produce fruits due to partial or total pollination failure at the turn of the century. The farmers had to resort to human pollination, where all family members — including children — used brushes to hand-pollinate the apple flowers. Now, apple orchards are gradually being replaced by plums, walnuts, and loquats as well as vegetables, which do not require human pollination.

Biodiversity research continues to unravel new aspects of pollinator-plant interactions. Accumulating field data can be very time consuming and it requires accurate species identification, which is now rarely taught in schools and universities. Thanks to websites and phone apps, it is now possible to use crowdsourcing as a means of gathering large volumes of data, by connecting beginners with experts. On www.ispotnature.org (currently there are four communities in UK and Ireland, Chile, Hong Kong and Southern Africa) photographs are posted for identification and over 94% eventually receive one. In January 2010, designated experts contributed to 60% of determinations; yet just four years later, 60% of observations were determined without input from experts.

Data will help to see how pollinator decline applies both nationally and regionally, ensuring we are better informed to find ways and means of redressing the balance. Pollinators are an essential part of the perpetuating — flower to seed to flower cycle — of the majority of plants on our planet.

The knock-on effect of a world without pollinators is inconceivable.

Right
Insect friendly flowers throughout the seasons, all in Surrey, UK.

Spring: A bumblebee forages for nectar on a female sallow or goat willow (*Salix caprea*) catkin.
Summer: Lavender attracts many butterflies, including small tortoiseshells (*Aglais urticae*).
Autumn: A female ivy bee (*Colletes hederae*) collects brood pollen from ivy (*Hedera helix*) — also an important autumn nectar source. This bee reached Britain in 2001 in Dorset.
Winter: Single snowdrops attract honeybees, bumblebees and hoverflies. As a hoverfly (*Eristalis tenax*) forages on nectar, pollen falls onto the underside of the insect.

BACK FROM THE BRINK

The mandrinette (*Hibiscus fragilis*) is a critically endangered shrub found on two Mauritius mountains and on Rodrigues. It is threatened by the invasion of the cultivated *Hibiscus rosa-sinensis* which competes and hybridises with it. Kew Gardens are propagating plants.

Clay's hibsicus (*Hibiscus clayi*) showing a conspicuous splayed out five-lobed stigma. This critically endangered species is endemic to Kauai Island, Hawaii and Kew Gardens are also propagating these plants.

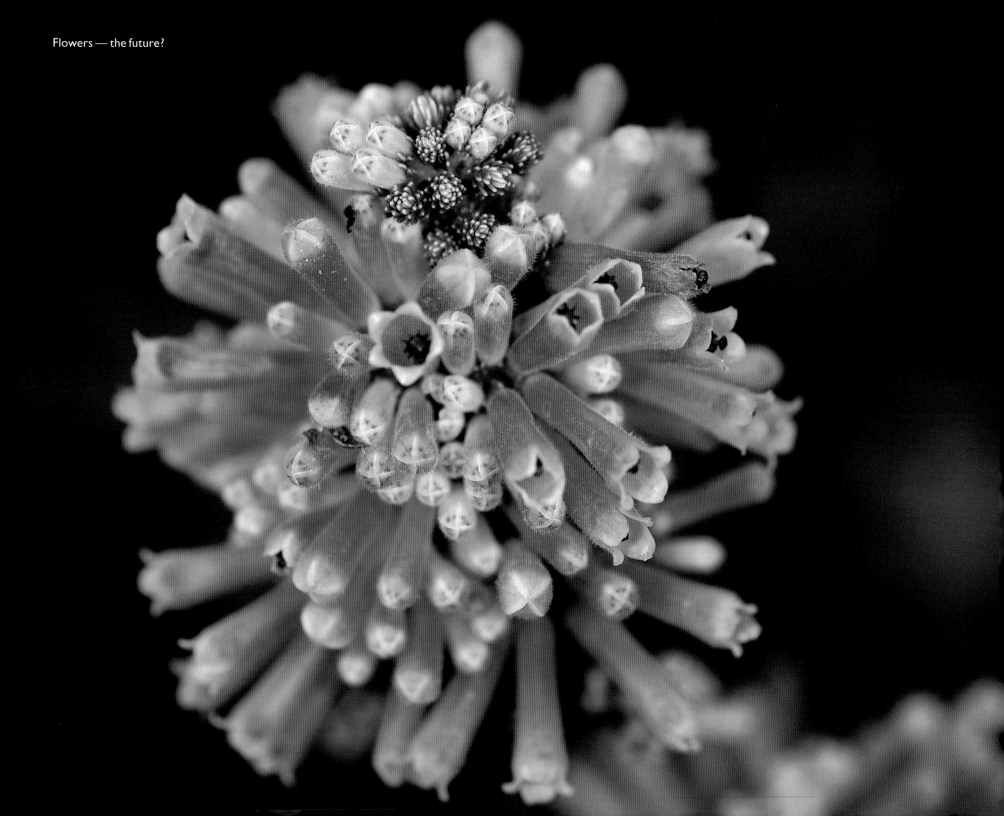

The pollinators of the tiny cocoa (*Theobroma cacao*) flowers (inset) was a mystery until the 1950's. At dawn, tiny (0.5–2mm) ceratopogonid midges pollinate the flowers, which eventually produce large golden cocoa pods. Kew Gardens, Surrey, UK.

A NATURAL HYBRID

The natural hybrid Hong Kong orchid tree (*Bauhinia blakeana*) is widely planted as a roadside tree. The parent trees share several floral visitors — including honeybees *Apis cerana* and *A. mellifera*, the bamboo carpenter bee (*Xylocopa iridipennis*) and the common mormon butterfly (*Papilio polytes*).

THE PARENTS

Guarianthe aurantiaca is an orange flowered orchid from Costa Rica and Mexico, and one parent of the naturally formed *Guarianthe × guatemalensis* hybrid. Kew Gardens, Surrey, UK.

The orchid *Guarianthe skinneri* is the national flower of Costa Rica, but also occurs in every country in Central America. This is the other parent of the naturally formed *Guarianthe × guatemalensis* hybrid. Kew Gardens, Surrey, UK.

THE NATURAL HYBRID

Left and above

Guarianthe × guatemalensis is a natural hybrid orchid of *Guarianthe auriantiaca × Guarianthe skinneri* found in Guatemala. Characteristics from both parents are visible. The colour of the blooms vary as shown by these two sprays, both from the same plant. Kew Gardens, Surrey, UK.

BREEDING FOR BEAUTY

Above
Chrysanthemums have been bred in China for thousands of years. This spider chrysanthemum cultivar has long narrow massed petals, without any stamens, so is of no benefit to insect visitors. At a show in Urumqi Botanical Garden, China.

Right
Another chrysanthemum variety on show in China, has the underside of the reflexed petals a contrasting colour to the top side. This cultivar is bred for beauty and does not attract insect visitors. Urumqi Botanical Garden, China.

Outside the exhibition, insects feast on potted chrysanthemums with small flowers
and simple outer ray petals. Painted lady butterflies, a silver Y moth and two hoverflies
feed on nectar from accessible disc florets. Urumqi Botanical Garden, China.

The open-bowl flower of the dog rose (*Rosa canina*) provides easy access to beetles, flies, bees and hoverflies. This beetle is feeding on exposed pollen. Hampshire, UK.

Rosa 'Warwick Castle' is a modern shrub rose with the fully double, fragrant, rich pink flowers that are attractive to gardeners; but the massed tightly packed petals make it an unattractive flower for insect visitors. Rose Garden, Warwick Castle, UK.

Left
The single open trumpets of wild daffodils
(*Narcissus pseudonarcissus*) are accessible to
insect visitors; here growing in profusion in the
churchyard at Farndale in Yorkshire, UK.

A double *Narcissus* cultivar with
extra petals and a double orange
trumpet has no stamens or
accessible nectar, so it does not
attract insects. Farnham, Surrey, UK.

AN INVASIVE ALIEN

Introduced Himalayan balsam (*Impatiens glandulifera*) grows vigorously and outcompetes smaller native plants; copious nectar lures bumblebees away from indigenous species. A bumblebee picks up pollen as it enters a flower. Farnham, Surrey, UK.

BACK TO NATURE
Disused railway cuttings can provide valuable native wildflower reserves in urban areas or through agricultural land. Here, fireweed, yellow loosestrife and ox-eye daisies flower in summer in Haskayne Cutting Nature Reserve, Lancashire, UK.

NATIVE FLORA ENHANCES ROADSIDES
Left
Lady Bird Johnson, a former First Lady, pioneered the planting of native wildflowers along Texas highways, notably prairie flowers like these bluebonnets (*Lupinus texensis*) and Indian paintbrush (*Castilleja indivisa*). Texas, USA.

A male southern double-collared sunbird (*Cinnyris chalybeus*) feeds on nectar from a pincushion protea (*Leucospermum cordifolium*) in Kirstenbosch NBG — the world's first botanic garden devoted to indigenous flora. The native Cape flora attracts a variety of local pollinators. South Africa.

A fallow field reverts back to a floriferous field with large poppies (*Glaucium grandiflorum*) and other wildflowers. The flowers attract pollinators and birds appreciate the seeds. Southern Turkey.

PHOTOGRAPHIC NOTES

Digital photography provides instant feedback of captured images, with the date, time and GPS retained in the metadata. No photo of an organism visiting a flower proves pollination has taken place; but if a visitor seen feeding on a flower moves to another of the same species, it may show pollen being transferred to the stigma. Most of the flower portraits are focus stacks that show anatomical details, while ultra violet flash reveals UV floral guides and the position of exposed nectar.

In the field

With a digital single lens reflex (D-SLR) camera, a 105mm or 200mm macro lens allows a greater working distance than shorter macro lenses, reducing the odds of insects taking evasive action. A macro flash unit is useful to boost low light levels. The first stage in the pollination story is to capture a clear view of the visitor making contact with exposed pollen. To show how an insect feeds, shoot either head or side on; the latter view will also show the colour of a pollen load in a bee's pollen basket.

Honeybees and bumblebees announce when they approach flowers by their buzzing sound. Silent visitors are spotted by scanning flowers and focusing in on darker mobile objects. Some systematically move up a spike while others move down. Butterflies tend to feed on the uppermost flower sprays. Longer lenses are necessary for frame-filling portraits of birds feeding on flowers. Wary species may even need a hide. In public gardens, birds are much more approachable, since they are habituated to people.

Working at night is challenging because our night vision is so poor. Focusing the camera is a problem unless an assistant shines an LED torch, covered with a red filter, onto the flower. Once insects begin to feed, most will tolerate a normal flash.

In the studio

Focus stacking

Most of the focus stacks were taken either at Kew or in my studio. With a deep funnelled flower such as the Chilean bellflower (*Lapageria rosea*) no amount of stopping down the lens to a small aperture will get everything pin sharp from the patterned corolla lobes, to the stamens and stigma half way down and the nectaries at the base of the funnel. The solution is to take a series of focus 'slices' by moving the camera forward in equidistant steps in a similar way as a CT body scan. The stack of partially focused images is then blended into a single focused image using software such as Helicon Focus or Zerene Stacker: http://www.heliconsoft.com or http://www.zerenesystems.com.

Essential factors for a successful focus stack are a static subject plus a constant light source and manual focus. The stem of the flower should be clamped (a Plamp is ideal). Mount the camera on a focusing rail, so it is moved forward in equal increments after manually focusing on the closest plane. The light can be window light, any continuous light source or flash. The total number of exposures will vary depending on the magnification used and the depth of field required.

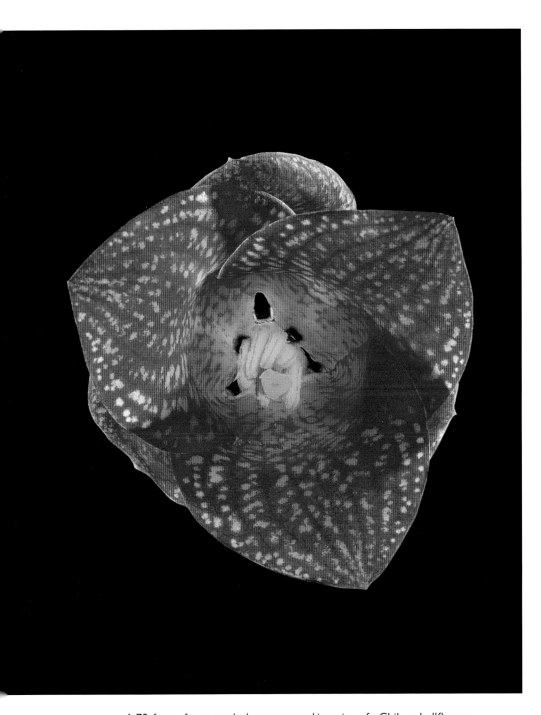

A 73-frame focus stack shows spotted interior of a Chilean bellflower (*Lapageria rosea*), six stamens surrounding the stigma and the dark lined nectaries in the base. Kew Gardens, Surrey, UK.

High speed flash

The only way to record a sudden explosive pollen release, without it appearing as a fuzzy blur, is to use high speed flash. Three Einstein 640 flash heads set to 1/13,000 second, froze the rivulet of pollen induced by a tuning fork simulating buzz pollination by bumblebees.

Seeing the unseen with UV

The typical human eye is only able to see wavelengths within the visible spectrum from about 400–750nm, so we can see the coloured floral guides present in many flowers, but not the 'hidden' UV guides. The UV spectrum has no distinct colours, which will vary depending on how UV images are captured and not least the colour space set on a digital camera and computer. Instead, tonal contrast becomes significant — especially in flowers, which to our eyes appear uni-toned. Protective goggles need to be worn when working with UV lights, since direct rays may damage eyes.

Small LED UV torches will reveal a UV pattern, but they are not as powerful as a UV flash with a dark UV transmitting filter (290–410nm), which I used. The latter generates a lot of heat, so that after a few exposures the glass UV filter has to be allowed to cool. So, UV focus stacks can take 20–30 minutes longer than a focus stack in normal light.

Exposed nectar in many flowers fluoresces when lit by a UV light source in a dark room. Almond (*Prunus amygdalus*) nectar fluoresces blue while pineapple lily (*Eucomis bicolor*) nectar fluoresces yellow.

SELECTED BIBLIOGRAPHY

Angel, H. (2012). *Exploring Natural China*. Evans Mitchell, London.

Arditti, J., Elliott, J., Kitching, I. J. & Wasserthal, L. T. (2012). 'Good Heavens what insect can suck it' — Charles Darwin, *Angraecum sesquipedale* and *Xanthopan morganii praedicta*. *Bot. J. Linn. Soc.* 169(3): 403–432

Barth, F. G. (1985). M. A. Bieberman-Thorson (transl.), *Insects and Flowers: The biology of a partnership*. Princeton University Press, Princeton.

Barwick, M. (2004). *Tropical & Subtropical Trees: A worldwide encyclopaedic guide*. Thames & Hudson, London.

Bown, D. (2000). *Aroids — Plants of the arum family*. 2nd edn. Timber Press, Portland.

Buchmann, S. L. & Nabhan, G. P. (1996). *The Forgotten Pollinators*. Shearwater Books, Washington DC.

Campbell-Culver, M. (2001). *The Origin of Plants*. Hodder Headline, London.

Chittka, L. (1996). Did bee color vision predate the evolution of flower color? *Naturwissenschaften* 83:136–138

Cox, P. & Hutchinson, P. (2008). *Seeds of Adventure: In search of plants*. Garden Art Press, Woodbridge, Suffolk.

Davis, A. P. (1999). *The Genus Galanthus*. Timber Press, Portland/Royal Botanic Gardens, Kew.

Edwards, M. & Jenner, M. (2005). *Field Guide to the Bumblebees of Great Britain and Ireland*. Ocelli Ltd, Eastbourne.

Elton, C.S., (2000). *The Ecology of Invasions by Animals and Plants*. University of Chicago Press, Chicago.

Endress, P. K. (1994). *Diversity and Evolutionary Biology of Tropical Flowers*. Cambridge University Press, Cambridge.

Faegri, K. & Van der Pijl, L. (1979). *The Principles of Pollination Ecology*. Pergamon Press, London.

Fayaz, A. (2004). *Encyclopedia of Tropical Plants: The identification and cultivation of over 3000 tropical plants*. Firefly Books, Canada.

Fleming, T. H. & Kress, W. J. (2013). *The Ornaments of Life: Coevolution and conservation in the tropics*. University of Chicago Press, Chicago.

Gardner, C. & Gardner, B. (2014). *Flora of the Silk Road*. I. B. Tauris, London.

Gibbons, B. (2011). *Wildflower Wonders of the World*. New Holland, London.

Gherardi. F. & Angiolini, C.(2007). Eradication and control of invasive species. In: EOLSS UNESCO. *Encyclopedia of life support systems: biodiversity conservation and habitat management* (vol. 2), pp. 274–302. Eolss Publishers, Oxford.

Goldblatt, P. & Manning, J. (2008). *The Iris Family: Natural history & classification*. Timber Press, Portland.

Goulson, D. (2013). *A Sting in the Tale*. Jonathan Cape, London.

Grant, K. A. & Grant, V. (1968). *Hummingbirds and their Flowers*. Columbia University Press, New York & London.

Grimshaw, J. (2002). *The Gardener's Atlas*. Firefly Books, Willowdale, Ontario.

Hansen, D. M., Olesen, J. M., Mione, T., Johnson, S. D., & Müller, C. B. (2007) Coloured nectar: distribution, ecology, and evolution of an enigmatic floral trait. *Biol. Rev.* 82 (1) 83–111

Harrap, A. & Harrap, S. (2005). *Orchids of Britain and Ireland*. A. & C. Black, London.

Heinrich, B. (1993). *The Hot-blooded Insects: Strategies and mechanisms of thermoregulation*. Harvard University Press, Cambridge, MA.

Heywood, V. H., Brummitt, R. K., Culham, A. & Seberg, O. (2007). *Flowering Plant Families of the World*. Royal Botanic Gardens, Kew.

Kirk, W. (2006). *A Colour Guide to Pollen Loads of the Honey Bee*. International Bee Research Association (IBRA), Cardiff.

Knuth, P. (1906–1909). *Handbook of Flower Pollination,* 3 volumes based on Hermann Müller's work, *The fertilisation of flowers by insects.* (trans.) J. R. Ainsworth Davis. Clarendon Press, Oxford.

Lee, D. (2007). *Nature's Palette.* University of Chicago Press, Chicago.

Lewington, A. (1990). *Plants for People.* Natural History Publications, London.

Lunau, K., (2007). Stamens and mimic stamens as components of floral colour patterns *Bot. Jahrb, Syst.* 127 (1) 13–41

Mabberley, D. J. (1997). 2nd edn. *The Plant-Book: A portable dictionary of vascular plants.* Cambridge University Press, Cambridge.

Nicolson, S. W., Nepi, M. & Pacini, E. (eds) (2007). *Nectaries and Nectar.* Springer, Dordrecht.

Phillips, R. & Rix, M. (1981). *Bulbs.* Pan Books, London.

Proctor, M., Yeo, P. & Lack, A. (1996). *The Natural History of Pollination.* Timber Press, Portland.

Silvertown, J., Harvey, M., Greenwood, R., Dodd, M., Rosewell, J., Rebelo, T., Ansine, J. & McConway, K. (2015). Crowdsourcing the identification of organisms: A case-study of iSpot. *ZooKeys* 480: 125–146. http://zookeys.pensoft.net/articles.php?id=4633

Warren, W. (1997). *Tropical Garden Plants.* Thames & Hudson, London.

Waser, N. M. & Ollerton, J. (eds.) (2006). *Plant-Pollinator Interactions from Specialization to Generalization.* University of Chicago Press, Chicago.

Willmer, P. (2011). *Pollination and floral ecology.* Princeton University Press, Princeton and Oxford.

Wilson, E. H. (1986). *A Naturalist in Western China.* Cadogan Books, London.

Yeo, P. F. (1993). *Secondary Pollen Presentation: Form, Function and Evolution* (*Pl. Syst. Evol.– Suppl. 6*). Springer-Verlag, Wien.

Useful websites

http://www.brc.ac.uk/plantatlas
Online atlas of British and Irish flora

http://www.bwars.com
Bees Wasps & Ants Recording Society (BWARS) is the national society dedicated to studying and recording bees, wasps & ants (aculeate Hymenoptera) in Britain & Ireland

http://darwin-online.org.uk
The world's largest and most widely used resource on Darwin

http://www.issg.org/database/welcome/
Global Invasive Database

http://www.iucnredlist.org/
The IUCN Red List of Threatened Species

http://www.kew.org/science-conservation/plants-fungi/species-browser?query&habitat=0&family=0&country=0
A plants and fungi species browser for information on almost 500 species on RBG Kew site

http://www.kew.org/science/tropamerica/neotropikey.htm
An interactive key and information resources for flowering plants of the Neotropics

http://plants.usda.gov
Database provides information on vascular plants and other US plants and its territories

http://www.proteaatlas.org.za
Information on proteas throughout southern Africa

http://www.tropicos.org
A botanical database with taxonomic information on plants, mainly from Central, and South America maintained by Missouri Botanical Garden

https://wiki.ceh.ac.uk/display/ukipi/Home
UK Insect Pollinators Initiative

http://www.wildflower.org/plants/result.php?id_plant=CHDU
NPIN: Native Plant Database, Lady Bird Johnson Wildflower Center

http://www.ywt.org.uk/sites/yorkshire.live.wt.precedenthost.co.uk/files/2011%2007%208038_churchyard_management_booklet_v2_0.pdf
Yorkshire Wildlife Trust, Yorkshire Living Churchyard Project

http://beekind.bumblebeeconservation.org

ACKNOWLEDGEMENTS

Many people helped in the production of this book. Most importantly, I greatly appreciate Richard Barley, Director of Horticulture, Royal Botanic Gardens, Kew, for granting me access to the Living Collections. Special thanks also go to all the Kew horticulturalists, both in the Gardens and the Nurseries, who tipped me off when certain plants were about to flower. Chris Clennett arranged my visit to Wakehurst Place to take *Fuchsia excorticata*. Several Kew scientists kindly looked at and discussed some of the images. Tremendous support was given to me by all the Kew Library staff, who managed to track down virtually all my requests.

I am indebted to all my guides who directed me to my overseas target species. Most especially, I thank Ian Green of http://www.greentours.co.uk who suggested locations for specific plants and guided me in Turkey and Mexico. Chris Gardner, a Greentours guide, played a huge part in getting me to the right place at the right time in Chile, Yunnan in China, Costa Rica, Tajikistan and Uzbekistan. Other Greentours guides were Paul Cardy in the Alpes-Maritimes, Dolomites and Sardinia, Mark Hanger in Western Australia, Vladimir Kolbintstev in Kazakhstan and Rosalind Slater in Tanzania. Mariana Delport of http://www.cape-ecotours.co.za arranged my tailor-made itineraries, drove and guided me in the Cape, South Africa in 2010 and in 2014 and Teresa Farino of http://www.iberianwildlife.com drove and guided me in Tenerife, the Canary Islands where I saw a Canary Island chiffchaff feeding on the sole *Isoplexis canariensis* in flower— within minutes of arriving.

Thanks also to the photographers from around the world who helped to fill some image gaps showing floral visitors on specific flowers. All are credited in the image captions.

The Bumblebee Conservation Trust and Jeremy Early kindly assisted with the identification of British bees; Graham A. Collins with British social wasps; Dr Dino J. Martins/Insect Committee of Nature Kenya identified the East African hoverfly and http://www.plantlife.org.uk suggested some useful resources for information on UK pollinators.

Special thanks go to http://www.advancedcameraservices.co.uk who converted several flash units so I could take the special UV images of flowers; to Kate Carter, who proofread all the copy and to my husband, Martin Angel, who has — as always — encouraged me throughout and accompanied me on some trips.

All plant names are as listed on the Plant List at the time of going to press. http://www.theplantlist.org/

BIOGRAPHY

HEATHER ANGEL worked as a marine biologist before becoming a peripatetic wildlife photographer. From intricate macro studies to action wildlife shots, her images combine scientific accuracy with a strong pictorial appeal.

Pollination Power portrays a small part of five years work Heather spent at Kew Gardens, in her studio and visiting 20 countries. During this time, Heather combined the research of others with her own field observations, which will appear in a more detailed book that looks at pollination studies on specific genera. The images here are a visual insight to the ways in which plants communicate with pollinators, showing the rewards on offer that ensure pollinators return to pick up pollen and carry it to another flower.

Acclaimed worldwide, Heather is a past President of the Royal Photographic Society and since 1994 she has been a Visiting Professor at Nottingham University. Major solo exhibitions of Heather's work have been on view in London, China, Malaysia, and Egypt. For more information about her work, see www.heatherangel.co.uk

Heather Angel rode on horseback to access the Aksu-Zhabagly Nature Reserve in Kazakhstan.